Kindle Fire Geekery

50 Insanely Cool Projects
for Your Amazon Tablet

Guy Hart-Davis

Mc
Graw
Hill

New York Chicago San Francisco Lisbon
London Madrid Mexico City Milan New Delhi
San Juan Seoul Singapore Sydney Toronto

The McGraw·Hill Companies

Cataloging-in-Publication Data is on file with the Library of Congress

Kindle Fire Geekery: 50 Insanely Cool Projects for Your Amazon Tablet

1 2 3 4 5 6 7 8 9 0   QFR QFR   1 0 9 8 7 6 5 4 3 2

ISBN   978-0-07-180273-4
MHID   0-07-180273-8

Sponsoring Editor
Roger Stewart

Editorial Supervisor
Janet Walden

Project Manager
Sapna Rastogi, Cenveo
Publisher Services

Acquisitions Coordinator
Ryan Willard

Copy Editor
Lisa McCoy

Proofreader
Carol Shields

Indexer
Jack Lewis

Production Supervisor
James Kussow

Composition
Cenveo Publisher Services

Illustration
Cenveo Publisher Services
and Lyssa Wald

Art Director, Cover
Jeff Weeks

Cover Designer
Ty Nowicki

This book is dedicated to Teddy.

About the Author

Guy Hart-Davis is the author of more than 80 computer books, including *iPhone Geekery; How to Do Everything: iPhone 4S; How to Do Everything: iPod touch; How to Do Everything: iPod & iTunes, Sixth Edition; The Healthy PC, Second Edition; PC QuickSteps, Second Edition; How to Do Everything with Microsoft Office Word 2007;* and *How to Do Everything with Microsoft Office Excel 2007.* His website is at www.ghdbooks.com.

Contents

Acknowledgments

I'd like to thank the following people for their help with this book:

- Roger Stewart for proposing and developing the book
- Ryan Willard for handling the acquisitions end
- Lisa McCoy for editing the manuscript with a light touch
- Janet Walden for assisting with the production of the book
- Sapna Rastogi for coordinating the production of the book
- Cenveo Publisher Services for laying out the pages
- Jack Lewis for creating the index

Introduction

Do you want to take your Kindle Fire to its limits—and then beyond them?

If so, this is the book for you.

This book shows you how to get the very most out of your Kindle Fire by using its built-in apps and features to the max—and adding more apps that will give you the extra features and functionality you crave.

What Does This Book Cover?

Here's what this book covers:

- Chapter 1, "Music Geekery," shows you how to get the most out of music on your Kindle Fire. First, you create top-quality music files from your CDs and put them on your Kindle Fire, then learn how to create digital files from your cassette tapes and records. You put your music on Amazon's Cloud Drive so you can stream it, then learn how to convert songs from file formats your Kindle Fire can't play to formats it can play, how to listen to the radio on your Kindle Fire, and how to play podcasts on the tablet. After that, we focus on using your Kindle Fire as your home stereo and your car stereo.

 DOUBLE GEEKERY

Why Is This Book Better than Other Kindle Fire Books?

Unlike other Kindle Fire books, this book assumes that you already know how to use your Kindle Fire—how to read books, browse the Web, install apps from the Appstore, and so on. (If you don't know all this, pick up a copy of the book *How to Do Everything: Kindle Fire,* also from McGraw-Hill, which covers this stuff in detail.)

Kindle Fire Geekery assumes you're already an intermediate or advanced Kindle Fire user—and that you want to become even more advanced. So it starts from that point, giving you a full book's worth of the good stuff you actually want, instead of grinding through all the basics you already know and then ending with a few pages of advanced material.

- Chapter 2, "Video Geekery," explores how to add TV shows to your Kindle Fire, how to stream movies from your computer to your Kindle Fire, and how to play a movie from your Kindle Fire on your computer. You then learn how to put your videos and DVDs on your Kindle Fire so you can play them anywhere. Finally, we look at how to use your Kindle Fire for in-car entertainment to keep your passengers happy while you battle the traffic on the freeway.

- Chapter 3, "Books Geekery," digs into ways to make the most of books on your Kindle Fire. I show you how to get the books you want without paying the earth for them, convert e-books from other formats into the Kindle Fire's preferred format, and make your Kindle Fire realize the e-books are books rather than documents. Toward the end of the chapter, you learn how to read your e-books in the bath safely, not to mention how to read your Kindle books on your computer—or on someone else's computer.

- Chapter 4, "System and Apps Geekery," teaches you to exploit the full potential of your Kindle Fire's system software and to install apps from sources other than Amazon's Appstore. You learn how to lock your Kindle Fire for privacy, update the operating system and apps, troubleshoot lockups and crashes, and make both your Kindle Fire and its Silk browser run faster. I also explain secrets for entering text more quickly in documents, how to customize the Carousel or replace it, and how to use your Kindle Fire to keep up with your social networks on Facebook, LinkedIn, and Twitter.

 Note paragraphs with this icon provide extra information. For example, Chapter 4 also shows you how to prevent your children from going online with your Kindle Fire, how to turn off One-Click Buy to spare your credit cards grief, and how to protect your Kindle Fire against malware.

- Chapter 5, "Kindle Fire at Work Geekery," shows you how to use your Kindle Fire as a work tool, either in your workplace or outside it. You learn to connect your Kindle Fire to an enterprise wireless network using digital certificates and how to make it communicate with Exchange Server systems. I show you how to manage your e-mail like a pro; how to get your contacts, calendars, tasks, and notes on your Kindle Fire; and how to create and edit Office documents on it.

 Tip paragraphs with this icon provide tips, tricks, hints, and workarounds. For example, you can even send text messages from your Kindle Fire—see Chapter 5 to learn how.

Conventions Used in This Book

To make its meaning clear without using far more words than necessary, this book uses a number of conventions, several of which are worth mentioning here:

- Note, Tip, and Caution paragraphs highlight information to draw it to your notice.

Caution paragraphs with this icon warn you of pitfalls and tell you how to avoid them.

- Double Geekery sidebars provide in-depth focus on important topics.
- The pipe character or vertical bar denotes choosing an item from a menu on the PC or Mac. For example, "choose File | Open" means that you should click the File menu and select the Open item on it. Use the keyboard, mouse, or a combination of the two as you wish.
- The ⌘ symbol represents the COMMAND key on the Mac—the key that bears the Apple symbol and the quad-infinity mark on most Mac keyboards.
- Most check boxes have two states: *selected* (with a check mark in them) and *cleared* (without a check mark in them). This book tells you to *select* a check box or *clear* a check box rather than "tap to place a check mark in the box" or "tap to remove the check mark from the box." Often, you'll be verifying the state of the check box, so it may already have the required setting—in which case, you don't need to tap at all.

1 Music Geekery

Even if you bought your Kindle Fire mostly for reading, you'll find it great for music and video as well, especially as it's small enough to put in your pocket or purse and carry with you wherever you go. The Kindle Fire's speakers are good enough for private listening, so the Kindle Fire works well as a personal audio player and portable radio. Or you can plug in your headphones and—insert your favorite health warning here—crank up the volume as loud as you enjoy.

In this chapter, I'll show you how to get the most out of music on your Kindle Fire. We'll start by creating top-quality music files from your CDs and put them on your Kindle Fire, then look at how to create digital files from your cassette tapes and records.

We'll then move along to getting serious amounts of your music on Amazon's Cloud Drive so that you can play it on your Kindle Fire via streaming. After that, I'll show you how to convert songs from file formats your Kindle Fire can't play to formats it can play, how to listen to the radio on your Kindle Fire, and how to play podcasts on it.

Lastly, with your Kindle Fire able to play all your songs and other audio, you may not need another audio player. So we'll go through ways to use your Kindle Fire as your home stereo and as your car stereo.

Let's get started.

Project 1: Create Top-Quality Music Files for Your Kindle Fire from Your CDs

Amazon would be delighted if you loaded your Kindle Fire with music by buying hundreds or thousands of songs from its MP3 Store, but if you have CDs, you should start by putting their songs on your Kindle Fire. This process is called "ripping" and involves not just copying the songs from the CDs but also encoding them in a different format.

The best application for ripping songs from CDs is iTunes, which Apple provides for free and which I'll show you how to use in this section. If you have a PC, you can download, install, and be up and running with iTunes within a few minutes. If you have a Mac, iTunes will already be installed, so you can get started even more quickly.

DOUBLE GEEKERY

Rip Songs Using Windows Media Player

If you have a PC rather than a Mac, and you prefer not to use iTunes, you can rip songs with Windows Media Player, which comes with almost all versions of Windows. (If your version of Windows doesn't include Windows Media Player, you can download Windows Media Player from the Microsoft website, www.microsoft.com/download/.)

Ripping songs with Windows Media Player is mostly straightforward except that Windows Media Player's default format for compressed files is Windows Media Audio, a Microsoft proprietary format your Kindle Fire can't play. So you need to set Windows Media Player to create MP3 files, which your Kindle Fire can play. Follow these steps:

1. Choose Start | All Programs | Windows Media Player.
2. If Windows Media Player displays the Welcome To Windows Media Player screen, choose the Recommended Settings option button and click the Finish button. (You can also select the Custom Settings option button and click the Next button if you want to choose custom settings.)
3. When you see the main Windows Media Player window, open the Organize menu and click Options to display the Options dialog box.
4. Click the Rip Music tab to bring it to the front of the Options dialog box, as shown here.
5. Look at the folder path shown in the Rip Music To This Location box. If you need to change the path, click the Change button, use the Browse For Folder dialog box to pick the folder, and then click the OK button.
6. In the Rip Settings box, choose settings for ripping:
   - Choose MP3 in the Format drop-down list.
   - Make sure the Rip CD Automatically check box is cleared.
   - Leave the Eject CD After Ripping check box cleared unless you're sure you want Windows Media Player to eject each disc automatically after ripping it. (This setting can cause surprises on a laptop—for example, if you put your coffee next to the optical drive.)
   - Drag the Audio Quality slider to a suitable position along the Smallest Size–Best Quality axis. The 128 Kbps bitrate gives you adequate quality and the smallest files, but you may prefer to go for higher quality and larger files.
7. Click the OK button to close the Options dialog box.
   You can now slip a CD into your PC's optical drive and start ripping.

In this section, we'll get iTunes set up on your computer, configure it to create top-quality music files, and then rip CDs with it.

Install iTunes on Your PC

To install iTunes on your PC, follow these steps:

1. Open your browser, go to the iTunes Download page on the Apple website (www.apple.com/itunes/download/), and then download the latest version of iTunes.

 The iTunes Download page encourages you to provide an e-mail address so that you can receive the New On iTunes newsletter, special iTunes offers, Apple news, and more. Unless you want to receive this information, be sure to clear the check boxes—in which case, you don't need to provide an e-mail address.

2. If Internet Explorer displays a File Download – Security Warning dialog box like the one shown here, verify that the name is iTunesSetup.exe. Then click the Run button.

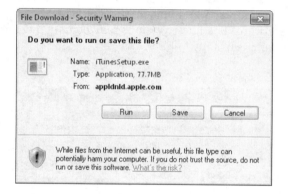

3. If Internet Explorer displays an Internet Explorer – Security Warning dialog box like the one shown here, verify that the program name is iTunes and the publisher is Apple Inc. Then click the Run button.

4. On the Welcome To iTunes screen, click the Next button.

5. On the License Agreement screen, read the license agreement, select the I Accept The Terms In The License Agreement option button if you want to proceed, and then click the Next button.

6. On the Installation Options screen (see Figure 1-1), choose installation options:

   - **Add iTunes And QuickTime Shortcuts To My Desktop**   Select this check box only if you need shortcuts on your desktop. The installation routine creates shortcuts on your Start menu anyway. The Start menu is usually the easiest way to launch iTunes. In Windows 7, you may want to pin iTunes to the taskbar so that you can launch it quickly.

   - **Use iTunes As The Default Player For Audio Files**   Select this check box if you plan to use iTunes as your main audio player. If you plan to use iTunes only for synchronizing the iPhone and use another player (for example, Windows Media Player) for music, don't make iTunes the default player. iTunes associates itself with the AAC, MP3, Apple Lossless Encoding, AIFF, and WAV file extensions.

   - **Automatically Update iTunes And Other Apple Software**   Select this check box if you want to set Apple Software Update to check automatically for updates to iTunes, QuickTime, Safari, and other Apple software you install. Apple Software Update is a utility that installs in Control Panel. Automatic updating is an easy way to make sure you have the latest versions of the Apple software. The latest versions may contain bug fixes or extra features, so having them is usually helpful.

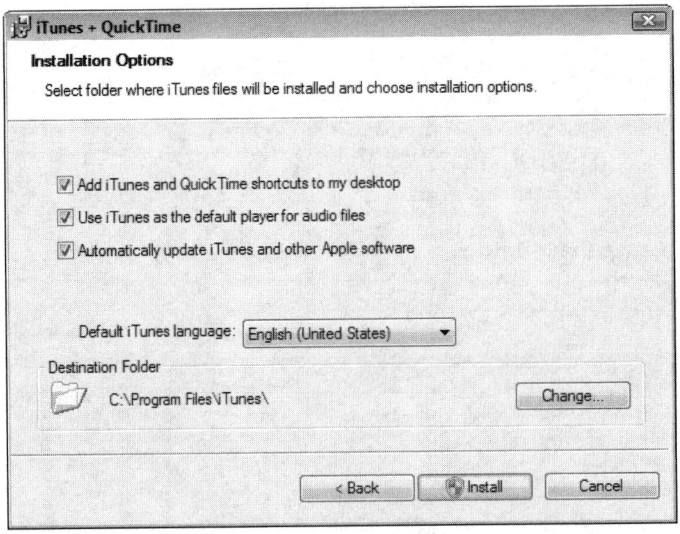

FIGURE 1-1   Choose whether to create shortcuts for iTunes and QuickTime on your desktop, use iTunes as the default audio player, and update iTunes and QuickTime automatically.

- **Default iTunes Language**   In this drop-down list, choose the language you want to use—for example, English (United States).
- **Destination Folder**   The installer installs iTunes in an iTunes folder in your Program Files folder by default. This is fine for most computers, but if you want to use a different folder, click the Change button, choose the folder, and then click the OK button.

7. Click the Install button to start the installation.

8. The installer displays the iTunes + QuickTime screen while it installs iTunes and QuickTime. On Windows XP, you need take no action until the Congratulations screen appears, telling you that iTunes and QuickTime have been successfully installed. But on Windows 7 and Windows Vista, you must go through several User Account Control prompts like the one shown here for different components of the iTunes installation (unless you've turned User Account Control off).

 On Windows 7 and Windows Vista, the User Account Control prompts may get stuck behind the iTunes + QuickTime screen. Look at the taskbar now and then to see if there's a flashing User Account Control prompt that you need to deal with before the installation can continue. Don't leave the installation unattended, because if you don't answer the User Account Control prompt, it times out and cancels the installation of the component.

9. When the Congratulations screen appears, leave the Open iTunes After The Installer Exits check box selected if you want to run iTunes immediately; otherwise, clear it. Then click the Finish button to close the installer.

Run iTunes for the First Time on Your PC

If you allowed the installer to run iTunes, the program now opens. If not, choose Start | iTunes | iTunes when you're ready to start running iTunes.

 The first time you launch iTunes, the program displays the iTunes Tutorial window, which contains tutorial videos showing you how to get started with iTunes and your iPhone. Watch the videos that interest you, and then click the Close button (the × button) to close the iTunes Tutorial window.

You then see the iTunes window with the Music item selected in the Source list. Because you haven't yet added any music to the iTunes library, the Music item displays information on downloading music, importing your CDs, and finding the music files in your home folder.

 If your computer is connected to the Internet, iTunes checks to see if an updated version of the program is available. If one is available, iTunes prompts you to download it (which may take a few minutes, depending on the speed of your Internet connection) and install it. After updating iTunes, you may need to restart your PC.

Launch iTunes on Your Mac

Apple installs iTunes on each Mac by default, so your Mac almost certainly contains iTunes.

To launch iTunes, click the iTunes icon on the Dock. If there's no iTunes icon on the Dock, click the Launchpad icon on the Dock, and then click the iTunes icon on the Launchpad screen.

 If you're using a version of Mac OS X before Mac OS X Lion, you don't have Launchpad. So if the iTunes icon doesn't appear on the Dock, launch iTunes from the Applications folder. Click the Finder icon on the Dock (or simply click the desktop) to activate the Finder, choose Go | Applications or press ⌘-SHIFT-A to open your Applications folder, and then double-click the iTunes icon. You can use this method on Lion and Mountain Lion as well, but Launchpad is easier.

Set Up iTunes for Ripping Your CDs

Using iTunes, you can rip your CDs at various different qualities called bitrates. The *bitrate* controls how much information iTunes includes in the files, which in turn controls their size. iTunes also lets you store the encoded files in different file formats, such as Advanced Audio Coding (AAC) and MP3; see the nearby sidebar for details.

 Because your Kindle Fire has only a modest amount of storage space, you'll need to keep down the size of music files as much as possible while retaining enough quality for the music to sound great.

To set up iTunes for ripping your CDs, follow these steps:

1. Display the iTunes dialog box or the Preferences dialog box:
   - In Windows, choose Edit | Preferences or press CTRL-COMMA or CTRL-Y to display the iTunes dialog box.
   - On the Mac, choose iTunes | Preferences or press ⌘-COMMA or ⌘-Y to display the Preferences dialog box.
2. If the General screen (see Figure 1-2) isn't displayed, click the General button on the tab bar to display it.

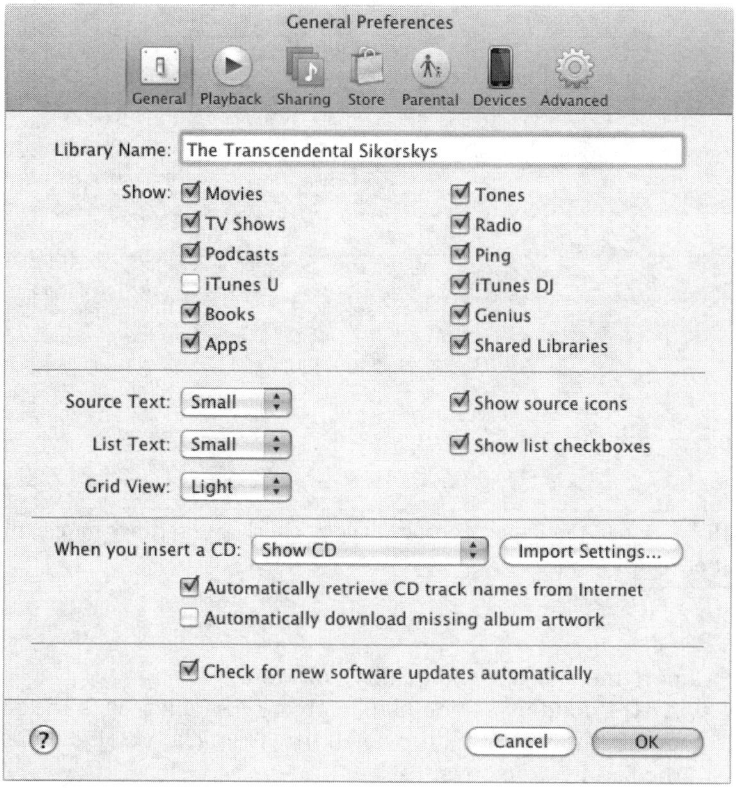

General Preferences

General Playback Sharing Store Parental Devices Advanced

Library Name: The Transcendental Sikorskys

Show: ☑ Movies   ☑ Tones
      ☑ TV Shows   ☑ Radio
      ☑ Podcasts   ☑ Ping
      ☐ iTunes U   ☑ iTunes DJ
      ☑ Books   ☑ Genius
      ☑ Apps   ☑ Shared Libraries

Source Text: [Small ⬍]   ☑ Show source icons

List Text: [Small ⬍]   ☑ Show list checkboxes

Grid View: [Light ⬍]

When you insert a CD: [Show CD ⬍]   (Import Settings...)

☑ Automatically retrieve CD track names from Internet
☐ Automatically download missing album artwork

☑ Check for new software updates automatically

(?)   (Cancel) (OK)

FIGURE 1-2   On the General screen, choose the Show CD item
in the When You Insert a CD drop-down list, then click the Import
Settings button to display the Import Settings dialog box.

3. In the When You Insert A CD drop-down list, choose the Show CD item.

The When You Insert A CD drop-down list on the General screen of the iTunes
dialog box or the Preferences dialog box also offers these choices: Begin Playing,
Ask To Import CD, Import CD, and Import CD And Eject. The Import CD And Eject
setting can be good when you need to rip many CDs quickly, but it doesn't let you
check that the tag information is correct. For this reason, I suggest choosing the
Show CD item instead and checking the CD information before you rip the CD.

4. Select the Automatically Retrieve CD Track Names From Internet check box.
5. Select the Automatically Download Missing Album Artwork check box.
6. Click the Import Settings button to display the Import Settings dialog box (shown
   in Figure 1-3 with settings chosen for AAC files).

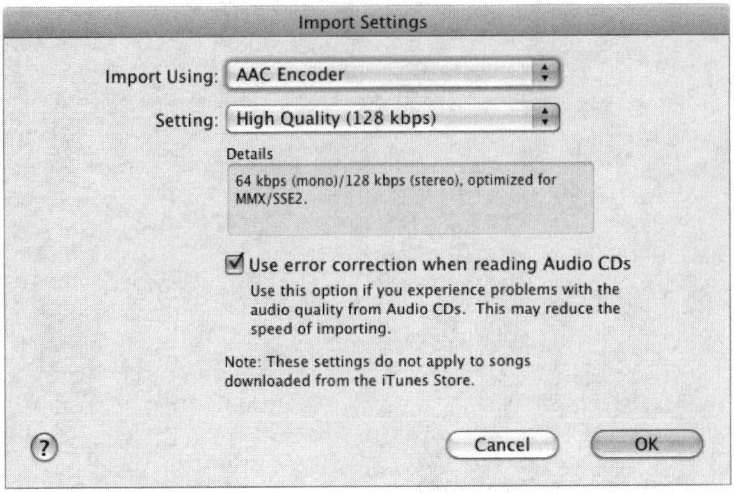

Import Settings

Import Using: | AAC Encoder |

Setting: | High Quality (128 kbps) |

Details

64 kbps (mono)/128 kbps (stereo), optimized for MMX/SSE2.

☑ Use error correction when reading Audio CDs

Use this option if you experience problems with the audio quality from Audio CDs. This may reduce the speed of importing.

Note: These settings do not apply to songs downloaded from the iTunes Store.

Cancel   OK

FIGURE 1-3 In the Import Settings dialog box, choose settings for creating files in the AAC format or the MP3 format.

7. Choose the encoder and quality you want:
   - **AAC Encoder** If you plan to create AAC files, click the AAC Encoder item in the Import Using drop-down list. Then choose High Quality (128 Kbps) in the Setting drop-down list.
   - **MP3 Encoder** If you want to create MP3 files, click the MP3 Encoder item in the Import Using drop-down list. Then choose High Quality (160 Kbps) in the Setting drop-down list.
8. Select the Use Error Correction When Reading Audio CDs check box.
9. Click the OK button to close the Import Settings dialog box.
10. Click the OK button to close the iTunes dialog box or the Preferences dialog box.

DOUBLE GEEKERY

Which Music File Formats Can Your Kindle Fire Play?

Amazon's MP3 Music Store sells song files for download in the MP3 format, just as its name suggests. So it's no surprise that the Kindle Fire can play MP3 files.

MP3 is technically a proprietary format, so you need to license a codec (*coder/decoder*) to create MP3 files. Because iTunes and other ripping programs such as Windows Media Player include a licensed codec, you don't need to worry about licensing a codec separately—so you can create MP3 files freely using programs such as iTunes and Windows Media Player.

Your Kindle Fire can also play files in these formats:

- **AAC**   Advanced Audio Coding (AAC) is Apple's preferred audio format and is the file format the iTunes Store uses for almost all of the songs it sells. AAC is technically superior to MP3, and most people find AAC files sound a little better than MP3 files at the same bitrate. Unless you have a good reason for preferring the MP3 format, AAC is your best choice for getting high music quality and lots of songs on your Kindle Fire.
- **OGG**   Ogg Vorbis (OGG) is an open-source format for compressed music files. Ogg Vorbis files are similar in quality to MP3 files. Ogg Vorbis works fine, but it isn't widely used. If you get Ogg Vorbis files, your Kindle Fire will play them just fine. Otherwise, you're probably better off creating MP3 files or AAC files than Ogg Vorbis files.
- **WAV**   WAV files are uncompressed audio files. Their size depends on their quality, but they're typically very large compared to MP3 files, AAC files, or OGG files. Your Kindle Fire can play WAV files, but they waste a lot of space, so I recommend not using them. I'll show you how to convert files from one format to another later in this project.

Rip a CD with iTunes

To rip a CD with iTunes, follow these steps:

1. Insert the CD in your computer's CD drive.
2. If iTunes doesn't automatically open (if it's not running) or become active (if it is running), launch it from the Start menu or taskbar (on Windows) or the Dock (on the Mac).
3. When iTunes detects the CD, it downloads the CD's details from the Gracenote Media Database on the Internet and then displays the track list (see Figure 1-4).

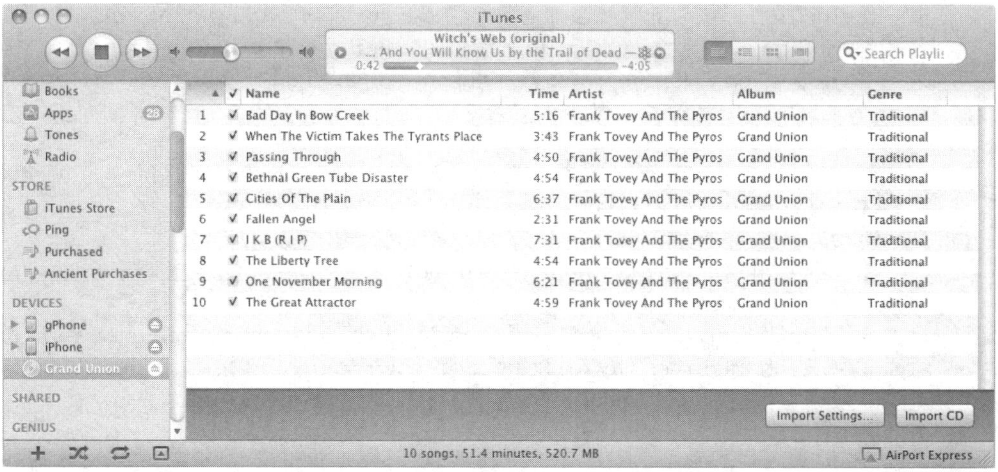

FIGURE 1-4   When iTunes displays the CD's track list, check that the details are correct and fix any spelling problems.

4. If you need to change the artist, album, genre, year, or disc number, or add the composer's name, right-click (or CTRL-click on the Mac) the CD's entry in the Devices category in the Source list, and then click Get Info on the context menu. In the CD Info dialog box that opens (shown here), change the information as needed, and then click the OK button.

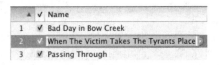

Select the Gapless Album check box in the CD Info dialog box if you want to rip the CD to songs that play back without any gap between them. This is good for live albums and concept albums. Select the Compilation CD check box only if the CD is a compilation from different artists; clear this check box if the CD is a compilation from a single artist.

5. If you need to change a single piece of information, click it twice, with a pause in between. iTunes displays an edit box around the information, as shown here. Edit the information or type a replacement, and then press ENTER (on Windows) or RETURN (on the Mac).

6. If you need to change multiple items of information, right-click (or CTRL-click on the Mac) the song, and then click Get Info on the context menu. iTunes displays the Item Information dialog box (see Figure 1-5). Add the information needed and then click the OK button to close the Item Information dialog box.
7. If you don't want to import any of the songs, clear its check box. (iTunes selects all the check boxes by default.)

You can quickly clear or select all the check boxes by CTRL-clicking (on Windows) or ⌘-clicking (on the Mac) one of them.

8. Click the Import button. iTunes imports the CD, showing its progress as it does so (see Figure 1-6).
9. When iTunes finishes ripping the CD, click the Eject button to the right of the CD's name in the Devices category of the Source list to eject the CD.

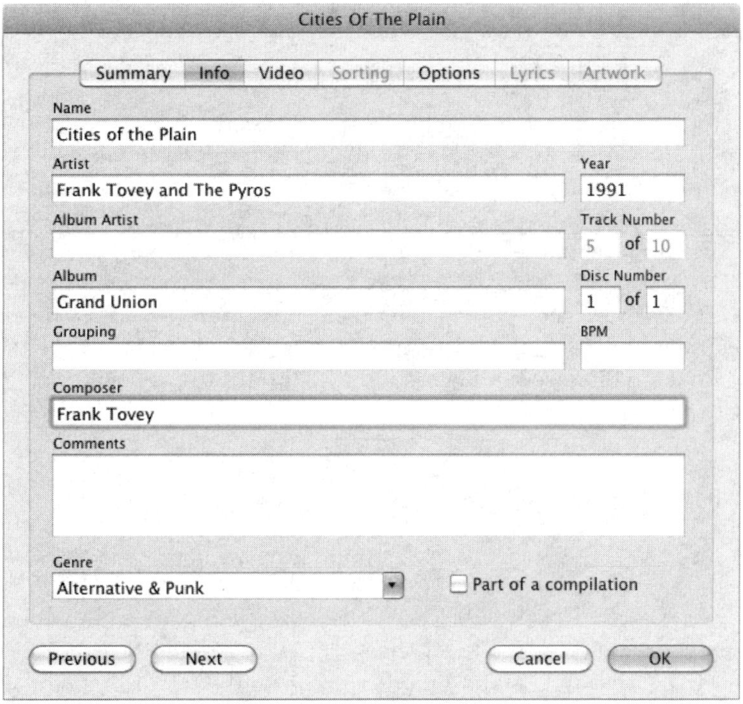

FIGURE 1-5 In the Item Information dialog box, you can correct the song's existing information or add various other pieces of information.

FIGURE 1-6 iTunes shows its progress as it rips the CD and creates files using the encoder you specified.

Convert a Song from One Format to Another Using iTunes

If you've got a song in a format such as Apple Lossless Encoding, WAV, or AIFF, you can quickly create an AAC version or MP3 version to put on your Kindle Fire.

In iTunes, right-click (or CTRL-click on the Mac) the song file, and then click the Create Version command on the context menu. This command shows your current import format in the name—for example, Create AAC Version or Create MP3 Version.

 If the Create Version command on the context menu shows a different format than you want to create, use the Import Settings dialog box to change the current encoder as discussed in the section "Set Up iTunes for Ripping Your CDs," earlier in this project.

Find the Song Files You've Ripped

After you rip the CD, the songs show up in iTunes, where you can play them easily and check that the audio quality is suitable.

To get at the actual song files, click one of the songs, and then choose File | Show In Explorer (on Windows) or File | Show In Finder (on the Mac). iTunes opens a Windows Explorer window or a Finder window to the folder that contains the files.

 You can also find the Show In Finder command on the context menu. Right-click (or CTRL-click on the Mac) a song in iTunes, and then click the Show In Finder item on the context menu.

Put the Ripped Files on Your Kindle

After opening a Windows Explorer window or a Finder window to the folder that contains the song files, you can quickly copy the files to your Kindle Fire. Follow these steps:

1. Connect your Kindle Fire to your computer as usual.
2. Open a Windows Explorer window or a Finder window showing the Kindle Fire's contents.
3. Double-click the Music folder to open it.
4. Open iTunes if it's not running, or activate iTunes if it is running.
5. Arrange the iTunes window so that you can also see the Music folder in the Windows Explorer window or Finder window.
6. In iTunes, select the songs you want to copy to your Kindle Fire.
7. Drag the songs from iTunes to the Music folder (see Figure 1-7), and then drop them there. Your computer copies the files across.

 Dragging between iTunes and your Kindle Fire works both ways: You can also copy songs from your Kindle Fire's Music folder to iTunes. Simply select the songs in the Music folder and then drag them to the Library section of the Source list in iTunes.

FIGURE 1-7   Drag songs from iTunes to your Kindle Fire's Music folder to copy the songs quickly and easily.

The next time you open the Music app on the Kindle Fire, you'll see the files you've added.

 You can't copy a playlist, an artist, or an album from iTunes to your Kindle Fire—just a group of songs. But you can easily select all the songs in an album, or all the songs by an artist, and drag them across.

Project 2: Transfer Your Tapes and Records to Your Kindle Fire

As you saw in the previous project, you can easily create AAC files or MP3 files from your CDs and put them on your Kindle Fire. But what if the songs you love are on tapes or records?

The good news is that you can still transfer these songs to your Kindle Fire. The bad news is that it takes a bit more effort than transferring CDs. First, you need to get the audio from its current analog format to a digital format such as MP3 or AAC. This means connecting a record player or cassette player to your computer and getting them to talk to each other. Second, you need to add the tag information to the files so that your computer and Kindle Fire can show you which file is which.

In this project, I show you how to

- Connect your sound source to your computer
- Download, install, and configure an application that can record high-quality audio
- Record the files
- Add tag information to the files

Connect Your Sound Source to Your Computer

First, you need to connect your sound source to your computer so that you can get the audio from the cassette tape or record to your audio-recording program.

If you have a working cassette deck or turntable, you can use a standard audio cable to connect it to your computer's audio input. For example, to connect a typical cassette player to a typical audio input, you'll need a cable with two RCA plugs at the cassette player's end (or at the receiver's end) and a male-end stereo miniplug at the other end to plug into your sound card. Figure 1-8 shows this type of cable. If the audio source has only a headphone socket or line-out socket for output, you'll need a miniplug at the source end too.

 Because record players produce a low volume of sound, you'll almost always need to put a record player's output through the Phono input of an amplifier before you can record it on your computer.

If your computer has a line-in port and a mic port, use the line-in port. If your computer has only a mic port, turn the source volume down to a minimum for the initial connection, because mic ports tend to be sensitive.

If you need to get a cassette deck or turntable for transferring your music, consider buying a USB-equipped cassette deck or turntable. Various companies make these, including Ion, Audio Technica, and Sony. Figure 1-9 shows such a turntable. Having a USB output gives you an easier way of getting a high-quality audio signal into your computer.

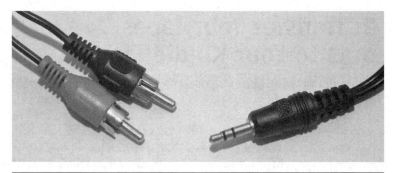

FIGURE 1-8 An RCA-to-miniplug cable is the most common means of connecting an audio source to your computer.

FIGURE 1-9   A cassette deck or turntable with a USB output is a great way to get high-quality audio into your computer.

Know Your Options for Recording Audio

Next, you need a suitable program for recording audio. I suggest you use Audacity, a free audio editor that's both well designed and easy to use.

 Both Windows and Mac OS X include programs that can record audio: Windows has Sound Recorder (which you'll find in the Accessories folder), and Mac OS X has QuickTime Player (which lurks in the Applications folder). On the Mac, if you have iLife, you can also use GarageBand to record audio. But Audacity has more features, which is why we'll use it in this project.

Usually you're better off getting a third-party program for recording your existing audio to create files for your computer and your Kindle Fire. Many audio-recording programs are available, but I recommend Audacity, which you can download for free from the Audacity webpage on SourceForge (http://audacity.sourceforge.net/). Audacity runs on both Windows and Mac OS X (and on Linux, if you use that too).

 If you want to try recording audio with Sound Recorder or QuickTime Player, visit my website (www.ghdbooks.com) for instructions.

Download and Install Audacity

Go to SourceForge (http://audacity.sourceforge.net) and download Audacity by following the link for the latest stable version and your operating system (for example, Windows or Mac OS X). At this writing, you download the file itself from one of various software-distribution sites by following a link from the Audacity page on SourceForge.

Once you've downloaded the Audacity distribution file, install it like this:

- **Windows** Double-click the distribution file, and then follow through the installation wizard that runs.
- **Mac OS X** Expand the downloaded file, and then drag the resulting Audacity folder to your Applications folder.

Run Audacity and Choose Your Language

The first time you run Audacity, you may need to choose the language you want to use—for example, English. You then see Audacity. Figure 1-10 shows the opening Audacity screen on the Mac, but the Windows version is almost identical.

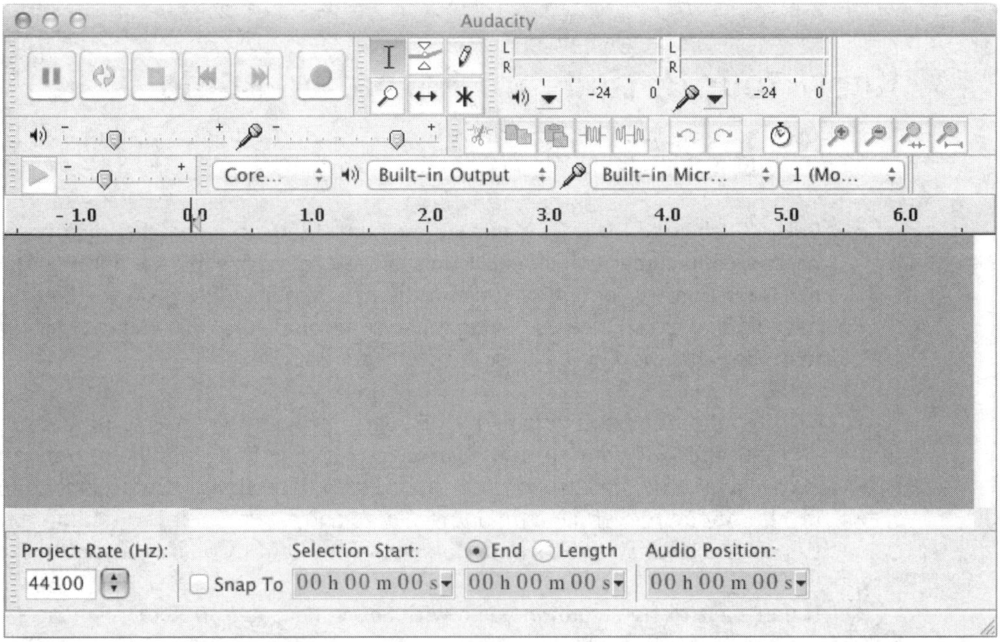

FIGURE 1-10 Audacity is a great freeware application for recording audio and fixing problems with it.

Install the LAME MP3 Encoder

After you install Audacity, you'll need to add an MP3 encoder if you want to be able to create MP3 files with Audacity. (Instead of adding the MP3 encoder, you can create WAV files with Audacity and then use iTunes to create MP3 files or AAC files.)

LAME is a recursive acronym that stands for LAME Ain't an MP3 Encoder.

Follow these steps to add an MP3 encoder to Audacity:

1. Open the Audacity Preferences dialog box in one of these ways:
   - **Windows** Press CTRL-P or choose Edit | Preferences.
   - **Mac OS X** Press ⌘-COMMA or choose Audacity | Preferences.
2. In the left pane, click the Libraries category to display its contents (see Figure 1-11).
3. Check the MP3 Export Library area. If the MP3 Library Version readout says "MP3 export library not found," you need to add an MP3 encoder.
4. Click the Download button in the MP3 Export Library area. Audacity opens a webpage in your default browser (for example, Internet Explorer on Windows or Safari on the Mac), giving instructions for downloading and installing the LAME MP3 encoder.
5. Follow the instructions to download the encoder and install it. When you have done so, its name and version appear next to the MP3 Library Version readout.

Leave the Audacity Preferences dialog box open so that you can set audio playback and recording settings, as discussed next.

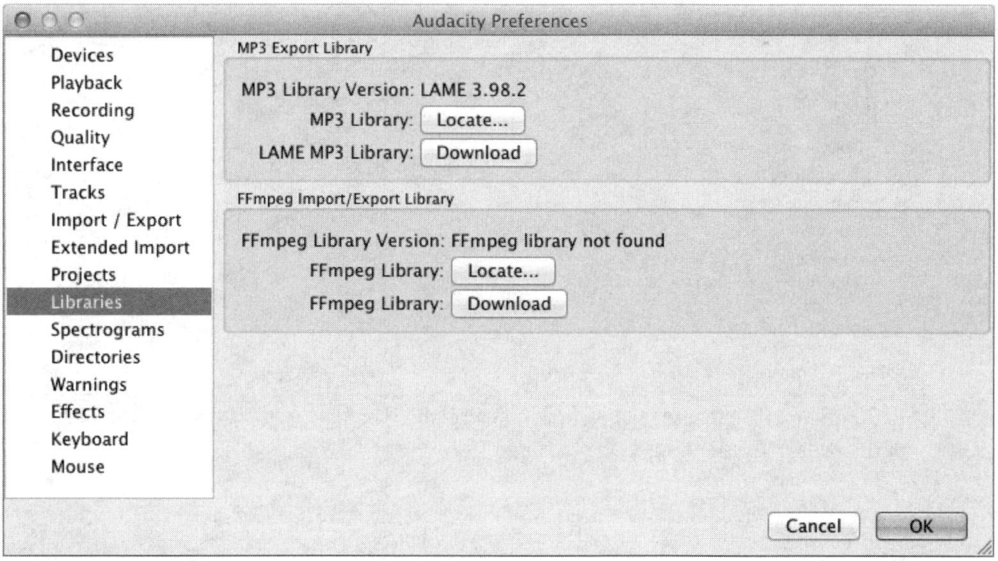

FIGURE 1-11 Use the File Formats tab of the Audacity Preferences dialog box to add an MP3 encoder to Audacity.

Choose Audio Playback and Recording Settings

With the Audacity Preferences dialog box still open, choose your audio playback and recording settings. Follow these steps:

1. In the left pane, click the Devices category to display its contents (see Figure 1-12).
2. In the Device drop-down list in the Playback box, choose the output device—for example, Built-in Output.
3. In the Device drop-down list in the Recording box, choose the device from which to record—for example, Built-in Input or a third-party device you've connected.
4. In the Channels drop-down list, choose 2 (Stereo).
5. Click the OK button to close the Audio Preferences dialog box.

Choose Your Audio Source

Now that you've got Audacity set up, you need to choose the audio source on your computer. In this section, we'll look at how to choose the audio source on Windows 7 and Windows Vista (together), Windows XP, and Mac OS X.

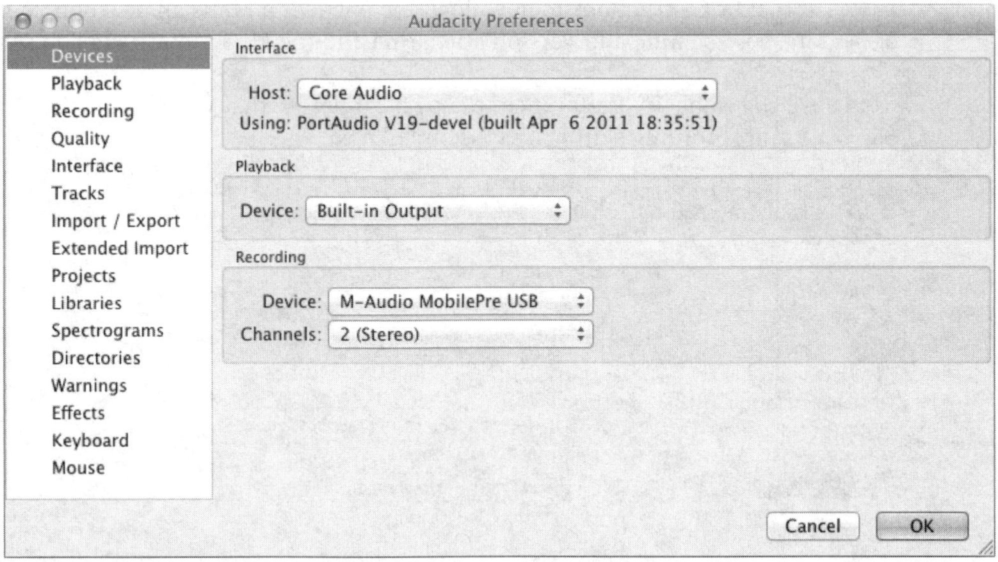

FIGURE 1-12 Choose your recording device and the number of channels in the Devices category of the Audacity Preferences dialog box.

Specify the Audio Source for Recording in Windows 7 or Windows Vista

To set Windows 7 or Windows Vista to accept input from the source so you can record from it, follow these steps:

1. Start your audio source playing so that you'll be able to check the volume level.
2. Right-click the Volume icon in the notification area, and then choose Recording Devices from the shortcut menu to display the Recording tab of the Sound dialog box (see Figure 1-13).
3. If the device you want to use is marked as Currently Unavailable, click the device, and then click the Set Default button. Windows makes the device available and moves the green circle with the white check mark from the other device to this device.
4. Verify that the signal level is suitable—for example, somewhere between the halfway point and the top of the scale for much of the input, as shown here.

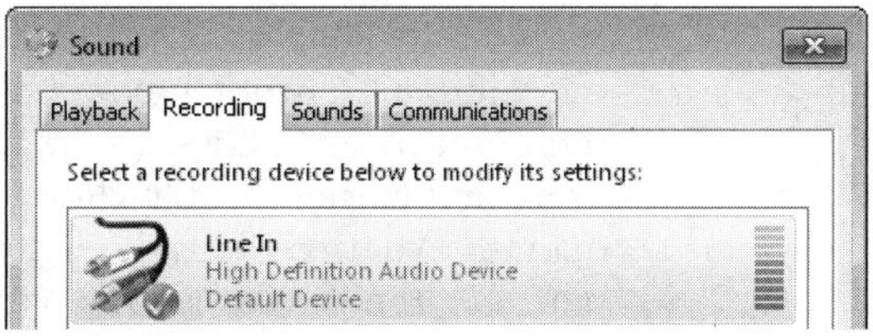

FIGURE 1-13   Before recording in Windows 7 or Windows Vista, you may need to change the recording device on the Recording tab of the Sound dialog box.

5. To change the recording level for the device, click the device in the list box, and then click the Properties button. In the Properties dialog box for the device, click the Levels tab to display its contents. You can then drag the input slider to the level needed. Click the OK button when you've finished.

 You can also change the left–right balance by clicking the Balance button and then dragging the L and R sliders in the Balance dialog box.

Specify the Audio Source for Recording in Windows XP

To set Windows XP to accept input from the source so you can record from it, follow these steps:

1. If the notification area includes a Volume icon, double-click this icon to display the Volume Control window. Otherwise, choose Start | Control Panel to display the Control Panel, click the Switch To Classic View link if it appears in the upper-left corner, double-click the Sounds And Audio Devices icon, and then open the Volume Control window from there. For example, click the Advanced button in the Device Volume group box on the Volume tab of the Sounds And Audio Devices Properties dialog box.

 Depending on your audio hardware and its drivers, the Volume Control window may have a different name (for example, Play Control).

2. Choose Options | Properties to display the Properties dialog box. Then select the Recording option button to display the list of devices for recording (as opposed to the devices for playback). The left screen in Figure 1-14 shows this list.

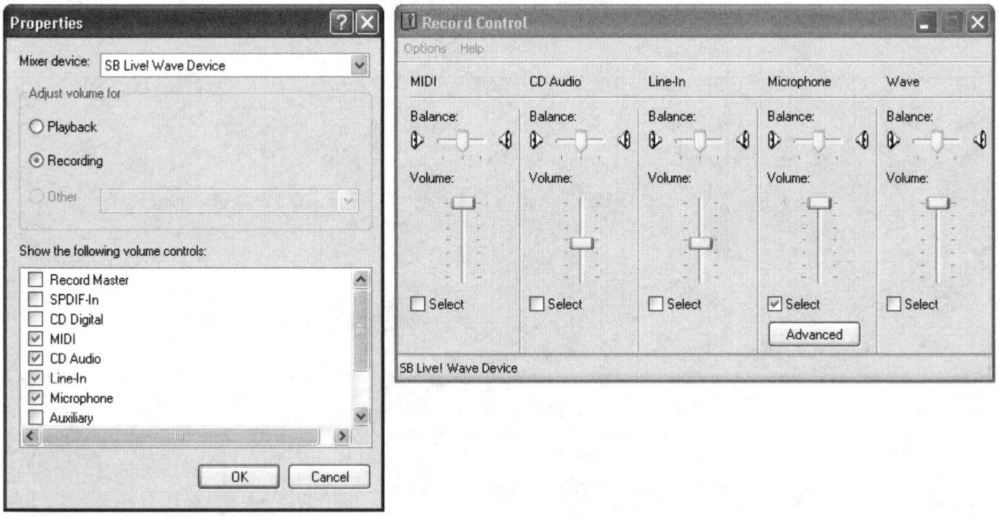

FIGURE 1-14 Click the Recording option button in the Properties dialog box (left) to display the Record Control window (right) instead of Volume Control.

3. Select the check box for the input device you want to use—for example, select the Line-In check box or the Microphone check box, depending on which you're using.
4. Click the OK button to close the Properties dialog box. Windows displays the Record Control window, an example of which is shown on the right in Figure 1-14. (Like the Volume Control window, this window may have a different name—for example, Recording Control.)
5. Select the Select check box for the source you want to use.
6. Leave the Record Control window open for the time being so you can adjust the input volume on the device if necessary.

Specify the Audio Source for Recording on the Mac

To specify the source on the Mac, follow these steps:

1. Choose Apple | System Preferences to display the System Preferences window.
2. In the Hardware section, click the Sound item to display the Sound preferences.
3. Click the Input tab to display it (see Figure 1-15).
4. In the Select A Device For Sound Input list box, select the device to use (for example, Line In).
5. Start some audio playing on the sound source. Make sure that it's representative of the loudest part of the audio you will record.
6. Watch the Input Level readout as you drag the Input Volume slider to a suitable level.

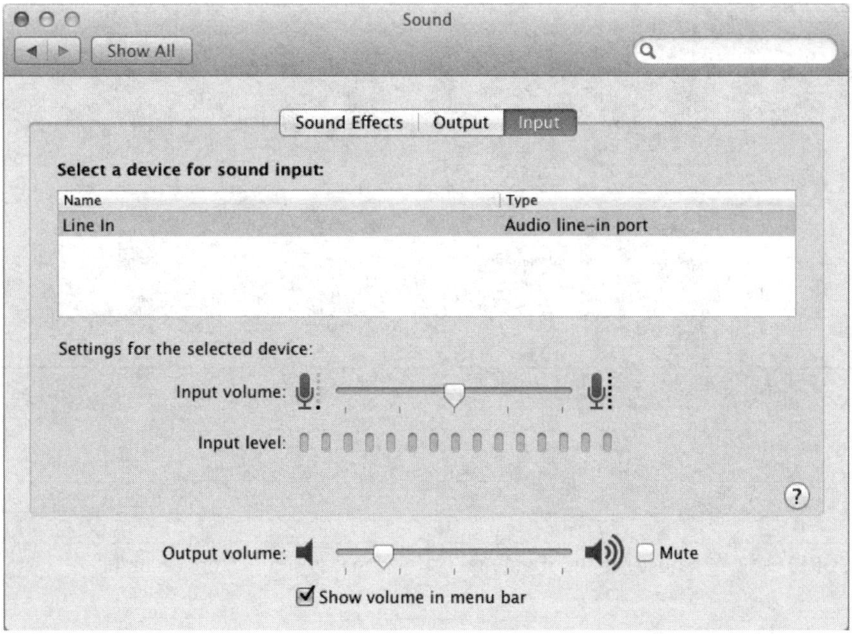

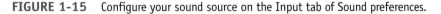

FIGURE 1-15 Configure your sound source on the Input tab of Sound preferences.

Record Audio with Audacity

To record audio with Audacity, follow these steps:

1. Start Audacity if it's not already running.
2. If necessary, choose a different sound source in the Default Input Source drop-down list on the right side of the window.
3. Cue your audio source.
4. Click the Record button (the button with the red circle) to start the recording (see Figure 1-16).
5. If necessary, change the recording volume by dragging the Input Volume slider (the slider with the microphone at its left end). When you've got it right, stop the recording, create a new file, and then restart the recording.
6. Click the Record button again to stop recording.
7. Choose File | Save Project to open the Save Project As dialog box, specify a filename and folder, and then click the Save button.

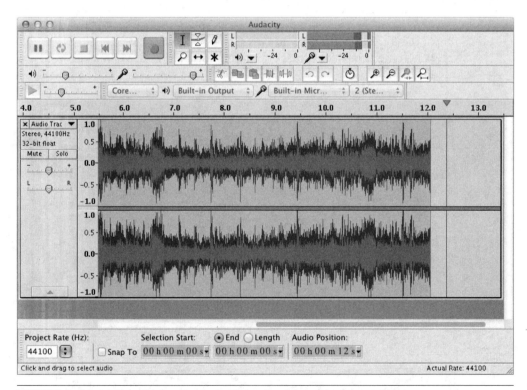

FIGURE 1-16   Adjust the input volume if the signal is too low or too high.

Remove Scratches and Hiss from Audio Files

If you record tracks from vinyl records, audio cassettes, or other analog sources, you may well get some clicks or pops, hiss, or background hum in the file. Scratches on a record can cause clicks and pops, audio cassettes tend to hiss (even with noise reduction such as Dolby), and record players or other machinery can add hum.

All these noises—very much part of the analog audio experience, and actually appreciated as such by some enthusiasts—tend to annoy people accustomed to digital audio. The good news is that you can remove many such noises by using the right software.

Unless you already have an audio editor that has noise-removal features, your best choice is probably Audacity. To remove noise from a recording using Audacity, follow these steps:

1. In Audacity, open the project containing the song.
2. Select a part of the recording with just noise—for example, the opening few seconds of silence (except for the stylus clicking and popping along).
3. Choose Effect | Noise Removal to open the Noise Removal dialog box (shown here ready for the second stage of the noise-removal process).
4. Click the Get Noise Profile button. The Noise Removal system analyzes your sample and applies the corresponding noise profile.
5. Select the part of the recording that you want to affect:
   - To affect the entire recording, choose Edit | Select | All. This is usually the easiest approach.
   - To affect only part of the recording, drag through it. For example, you might want to affect only the end of the recording if this is where all the scratches occur.
6. Choose Effect | Noise Removal to open the Noise Removal dialog box again.
7. Drag the sliders in the Step 2 box to specify how much noise you want to remove, the sensitivity, the frequency smoothing, and the attack and decay time.
8. In the Noise area, make sure the Remove option button is selected rather than the Isolate option button.
9. Click the Preview button to get a preview of the effect this will have, and adjust the sliders as needed.
10. When you're satisfied with the effect, click the OK button to remove the noise.
11. Choose File | Save to save your project.

Export the Recorded Files to MP3 and Put Them on Your Kindle Fire

When you are ready to export the audio file you've recorded, follow these steps:

1. Choose File | Export to display the Save As dialog box (shown here).

> Save As: Heaven in a Graveyard
> Where: 📁 Audacity
> Format: MP3 Files   Options...
> Cancel   Save

2. Choose MP3 Files in the Format drop-down list.
3. Click the Options button to display the Specify MP3 Options dialog box (shown here).

> **Specify MP3 Options**
> MP3 Export Setup
> Bit Rate Mode: ○ Preset ○ Variable ○ Average ● Constant
> Quality 320 kbps
> Variable Speed: Fast
> Channel Mode: ○ Joint Stereo ● Stereo
> Cancel   OK

4. In the Bit Rate Mode area, choose the bitrate mode by selecting the Preset option button, the Variable option button, the Average option button, or the Constant option button. A variable bitrate gives you the best quality for a given file size, but a constant bitrate is more compatible with older players.
5. In the Quality drop-down list, choose the bitrate you want to use—for example, 320 Kbps for top quality, or 128 Kbps if you need to pack as many songs as possible on your Kindle Fire.
6. If you selected the Variable option button, choose the variable speed in the Variable Speed drop-down list.
7. In the Channel Mode area, select the Joint Stereo option button or the Stereo option button, as needed. Normal stereo (the Stereo option button here) gives better sound at higher bitrates.
8. Click the OK button. Audacity returns you to the Save As dialog box.
9. Click the Save button. Audacity displays the Edit Metadata dialog box (see Figure 1-17).
10. Enter the song's metadata—the artist name, song name, album title, and so on.
11. Click the OK button. Audacity exports the song to an MP3 file.

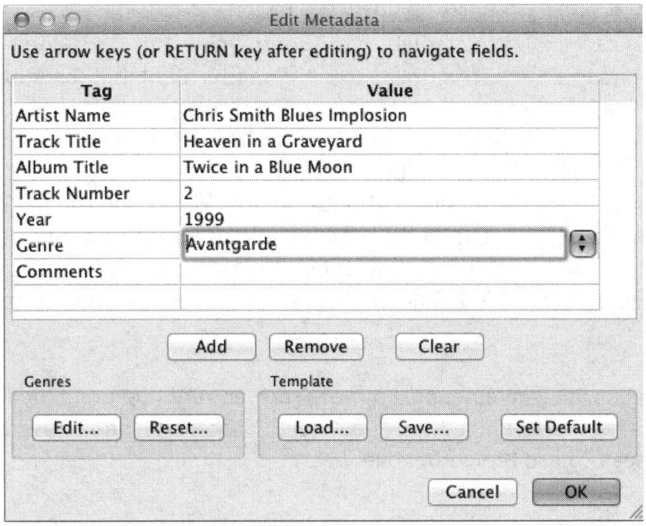

Edit Metadata

Use arrow keys (or RETURN key after editing) to navigate fields.

Tag	Value
Artist Name	Chris Smith Blues Implosion
Track Title	Heaven in a Graveyard
Album Title	Twice in a Blue Moon
Track Number	2
Year	1999
Genre	Avantgarde
Comments	

Add   Remove   Clear

Genres   Template

Edit...   Reset...   Load...   Save...   Set Default

Cancel   OK

FIGURE 1-17   Type the tag information for the song file in the Edit Metadata dialog box.

After creating the MP3 file, you can add it to your Kindle Fire by dragging it from a Windows Explorer window or a Finder window to your Kindle Fire. You can also add the MP3 file to iTunes and then add it to your Kindle Fire from iTunes as discussed in Project 4.

Project 3: Put Your Music on Amazon's Cloud Drive

As you know, your Kindle Fire has a nominal total capacity of 8GB, of which around 6GB is free for your media files. Even if you encode your songs at a moderate bitrate such as 128 Kbps, and you put only a modest amount of video on your Kindle Fire (as discussed in Chapter 2), you'll still be able to put only some of your music library on the Kindle Fire at a time.

The good news is that Amazon has a solution for you: Cloud Drive.

Understand What Cloud Drive Is

Cloud Drive is online storage that Amazon provides—space you can use to store files on the Internet so you can access them from any computer or device that has an Internet connection.

Amazon gives you 5GB of free storage on Cloud Drive with your Kindle account. That's only a modest amount, but you can buy more space if you need it. You may not need to buy more space because Amazon has another trick up its sleeve: Any songs you buy from Amazon are available on Cloud Drive, but they don't take up any of your storage space. So your storage space remains available for your own songs—those from your CDs or other media, or that you buy from sources other than Amazon.

You can put any files onto your space on Cloud Drive—for example, your current documents, videos, and photos. But Amazon provides a special tool called Cloud Player for working with music on Cloud Drive. Cloud Player gives you access to a tool called Amazon MP3 Uploader, which runs on both Windows and the Mac, and lets you easily upload MP3 and AAC files. If you use the MP3 and AAC formats (as I recommend), Amazon MP3 Uploader is the easiest way to put your songs on Cloud Drive.

 Don't put uncompressed audio files (such as WAV files) on Cloud Drive. The files will take ages to upload, will waste your space, and won't have the tag information you need to use them effectively.

You can access Cloud Drive from up to eight devices—for example, your Kindle Fire and several computers, or all your family's Kindles and computers.

Connect to Cloud Drive, Set It Up, and Get Amazon MP3 Uploader

To connect to Cloud Drive, follow these steps on either Windows or the Mac:

1. Open your web browser and go to the main Amazon.com website, www.amazon.com.
2. In the Shop All Departments box on the left, click the Amazon Cloud Drive link.
3. On the submenu that opens, click the Your Cloud Drive link. Your browser displays the Sign In page (shown here).

amazon.com   Your Account | Help

Sign In

What is your e-mail address?

My e-mail address is: [   ]

Do you have an Amazon.com password?

○ **No, I am a new customer.**

● **Yes, I have a password:** [   ]

Forgot your password?

[Sign in using our secure server ▶]

4. Type your e-mail address and password, and then click the Sign In Using Our Secure Server button. Your browser displays the Welcome To Cloud Drive page (see Figure 1-18).

5. Click the Launch Cloud Player link in the upper-left corner of the window to launch Cloud Player.

The first time you log in to Cloud Player, you must type the letters in a captcha image to prove you're human and select a check box saying you accept the terms of use for Cloud Player and Cloud Drive.

6. When you see the Cloud Player screen shown in Figure 1-19, click the Upload Your Music button in the upper-left corner to start uploading your music.

If Cloud Player displays a message saying that you must install the latest version of Adobe Flash Player, follow the instructions to do so. Cloud Player relies on Flash Player for some of its functionality, so don't dismiss the update notice as an irritation.

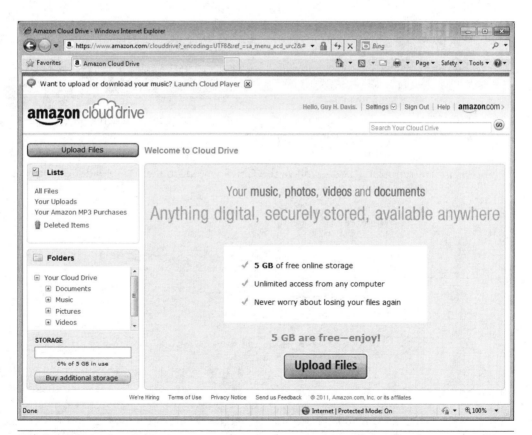

FIGURE 1-18   Launch Cloud Player by clicking the Launch Cloud Player link in the upper-left corner of the Welcome To Cloud Drive page.

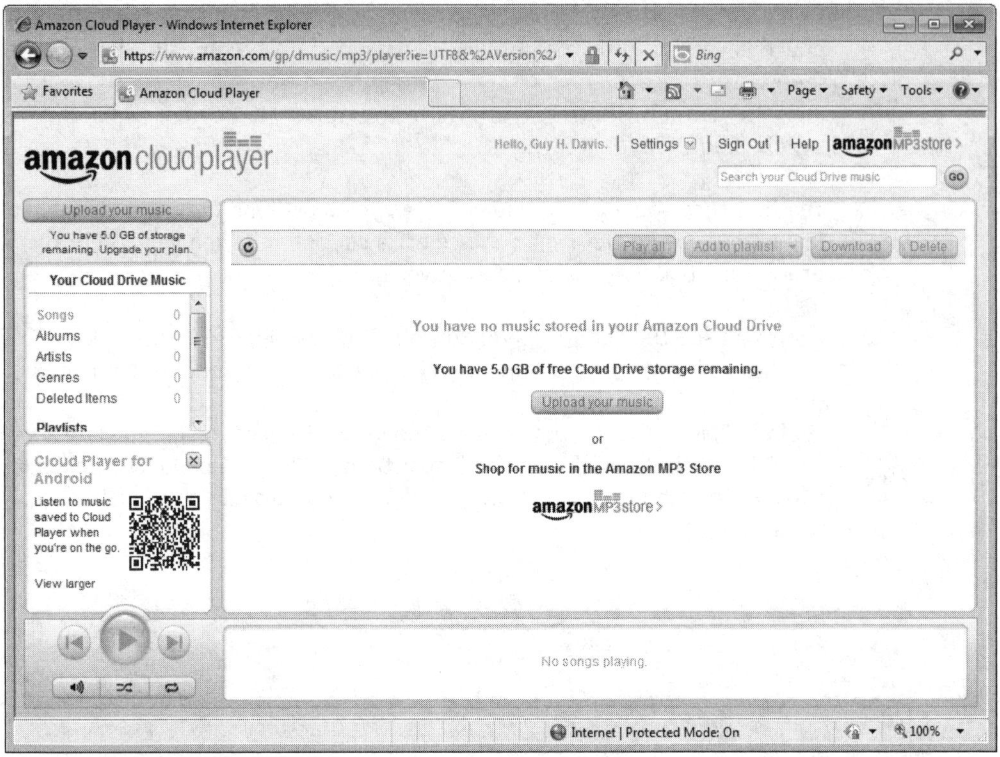

FIGURE 1-19 Click the Upload Your Music button in the upper-left corner of the Cloud Player screen to start uploading your music to Cloud Drive.

7. Your browser then displays the Get The Amazon MP3 Uploader dialog box (shown here).

8. Click the Download Now button, and then follow through the process of installing Amazon MP3 Uploader, as explained in the following subsections.

Install Amazon MP3 Uploader on Windows

When you click the Download Now button on Windows, Internet Explorer displays the File Download – Security Warning dialog box. Click the Run button to allow the file to run when the download completes.

Next, you'll see the Application Install wizard screen (see Figure 1-20). Here you can choose settings for the installation:

- **Install Adobe AIR**   Adobe AIR is required, so you can't clear this check box. AIR stands for Adobe Integrated Runtime, an Adobe technology for building applications.
- **Start Application After Installation**   Clear this check box if you don't want the installer to run Amazon Uploader when the installation finishes.

Click the Continue button. The Application Install wizard displays a screen showing the license agreement for Adobe AIR.

Click the I Agree button if you want to proceed with the installation, and then follow through the rest of the installation.

Install Amazon MP3 Uploader on the Mac

After downloading the Install Amazon MP3 Uploader disk image, mount the file if OS X doesn't automatically mount it for you. Usually, the easiest way to mount the file is to click the Downloads button on the Dock, and then click the icon for the Install Amazon MP3 Uploader disk image.

FIGURE 1-20   Choose settings for the Amazon MP3 Uploader on this screen of the Application Install wizard, and then click the Continue button.

In the Finder window that opens, double-click the Install Amazon MP3 Uploader icon to launch the installer.

 If the installer prompts you to get a new version of Adobe AIR, click the link to go to the Adobe AIR webpage. Click the Downloads link at the top, and then click the Download Adobe AIR button. When the download completes, install Adobe AIR, and then continue installing Amazon MP3 Uploader.

Next, you'll see the Application Install screen (see Figure 1-21). Here you can choose settings for the installation:

- **Install Adobe AIR**   Adobe AIR is required, so you can't clear this check box.
- **Start Application After Installation**   Clear this check box if you don't want the installer to run Amazon Uploader when the installation finishes.
- **Installation Location**   This text box shows the folder in which the installer will put Amazon MP3 Uploader—by default, /Applications/Amazon/Utilities/. If necessary, change to a different folder either by typing its path or by clicking the button to the right of the text box and then using the Open dialog box to select the folder.

Click the Continue button when you've made your choices, and then follow through the rest of the installation. You'll need to authenticate yourself as an Administrator as usual.

![Application Install screen showing Amazon MP3 Uploader installation preferences with Install Adobe AIR 3.1 and Start application after installation check boxes, an Installation Location field set to /Applications/Amazon/Utilities, and Cancel and Continue buttons.]

FIGURE 1-21   On the Mac, choose settings for the Amazon MP3 Uploader on this Application Install screen, and then click the Continue button.

Upload MP3 and AAC Files with Amazon MP3 Uploader

After installing Amazon MP3 Uploader, you can upload files to Cloud Drive with it. Follow these steps:

1. Launch Amazon MP3 Uploader if it's not already running:
   - **Windows**   Choose Start | All Programs | Amazon MP3 Uploader. Or, if you let the installer create a desktop shortcut for Amazon MP3 Uploader, double-click that shortcut.
   - **Mac**   Click the Launchpad button on the Dock, and then click the Amazon MP3 Uploader icon.
2. Click the Amazon Cloud Player link, and then sign in to Amazon if your browser isn't set to sign in automatically.
3. Click the Upload Your Music button to activate Amazon MP3 Uploader; then wait while Amazon MP3 Uploader scans your computer for songs and playlists.
4. Select the songs and playlists you want to upload. Expand the All Music listing as shown in Figure 1-22, and then clear the check box for any artist, album, song, or playlist you don't want to upload.
5. Click the Start Upload button. The lower part of the Amazon MP3 Uploader window shows you how the upload is progressing, together with the estimated time

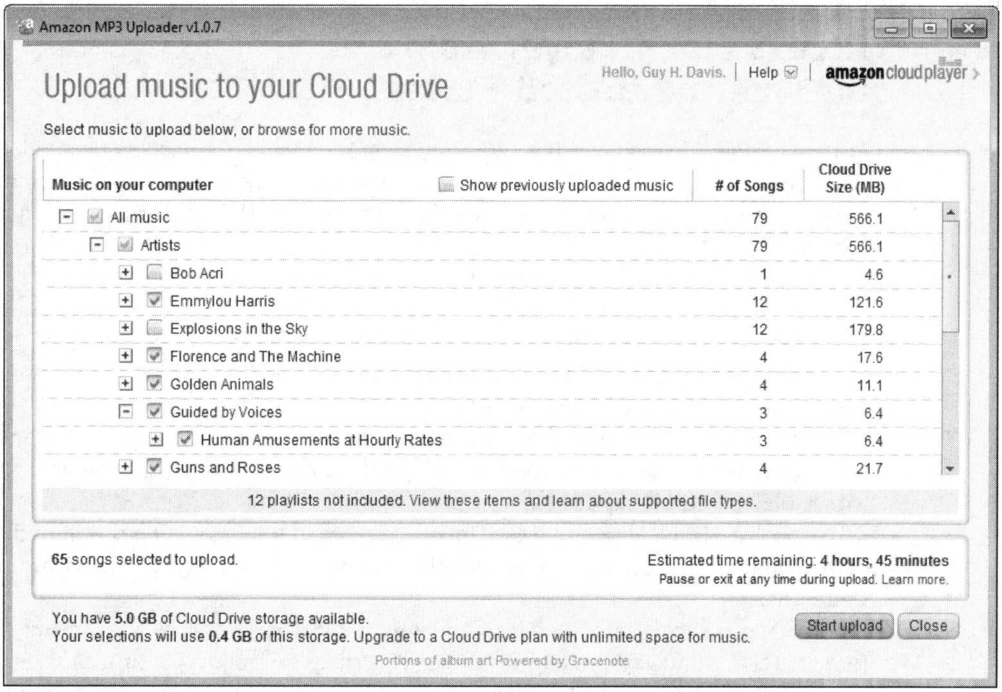

FIGURE 1-22   On the Upload Music To Your Cloud Drive screen, clear the check box for any item you don't want to upload, and then click the Start Upload button.

left, as shown here. You can pause the upload at any time by clicking the Pause Upload button (which replaces the Start Upload button) and resume uploading by clicking the Resume button (which replaces the Pause Upload button).

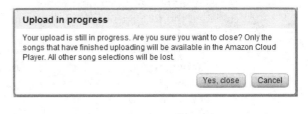

Click the Close button when you finish using Amazon MP3 Uploader. If an upload is running, or if you've paused an upload, Amazon MP3 Uploader displays the Upload In Progress dialog box (shown here) asking if you want to close Amazon MP3 Uploader and lose the song selections you haven't yet uploaded. Click the Yes, Close button if you do want to close Amazon MP3 Uploader.

Upload in progress

Your upload is still in progress. Are you sure you want to close? Only the songs that have finished uploading will be available in the Amazon Cloud Player. All other song selections will be lost.

Yes, close   Cancel

DOUBLE GEEKERY

Access Cloud Player from Outside the United States

At this writing, Cloud Player is a U.S.-only service. If you try to access Cloud Player from outside the United States, you'll see a message saying that "it appears you are attempting to use Amazon Cloud Player from outside the U.S." You then can't get any further.

Amazon uses the Internet Protocol (IP) address from which your computer is connecting to identify your geographic location. This IP address is normally the IP address of your network's router rather than the computer's own IP address, because your computer connects through the router. The location this gives is only approximate, but it's clear enough whether you're in the United States or not.

If you're outside the United States (for example, if you're traveling), you won't be able to use Cloud Player even though you log in using your usual Amazon account with a U.S.-based billing address.

If you need to be able to access Cloud Player when you're outside the United States, you need to use a proxy service that provides an IP address from within the United States.

If you need such a proxy service frequently, look at pay services such as Hide My Ass! (http://hidemyass.com) or Strong VPN (http://strongvpn.com). Despite its name, Hide My Ass! is a professional-grade service that provides reliable proxying for $6 to $12 a month, depending on how many months you pay for. Strong VPN provides similar services, but at this writing gets slightly fewer recommendations.

If you need proxying only occasionally and don't mind configuring your web browser manually, you can try searching for "open proxy server" and entering the resulting IP addresses and ports in your web browser. But you'll need to stay on your toes, because while most free open proxy servers provide redirection with good intentions, others have lower motives and try to grab passwords and other potentially valuable information that you send through them.

Upload Other Files to Cloud Drive

As you know, this project focuses on music, but you can also upload other files to Cloud Drive. To do so, just take your web browser to www.amazon.com, click the Amazon Cloud Drive link in the Shop All Departments box on the left, and then click the Your Cloud Drive link. After you sign in, you can manage your files and folders on the Your Cloud Drive screen (see Figure 1-23).

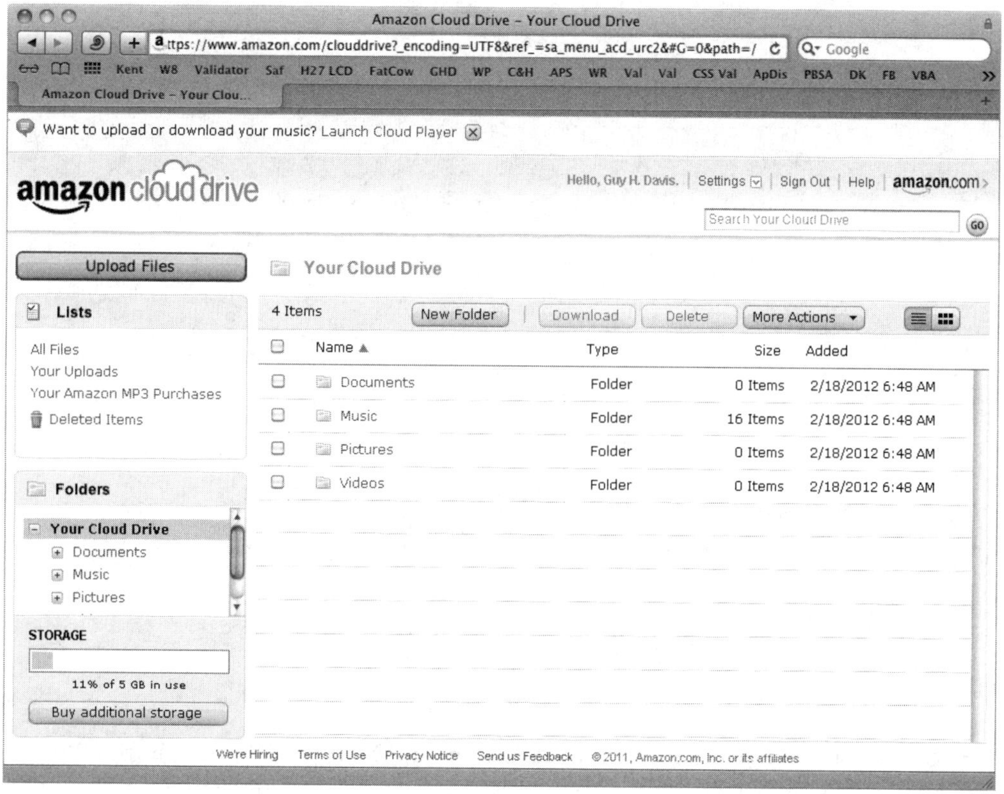

FIGURE 1-23   You can easily upload other files to the folders on your Cloud Drive.

To upload a file, follow these steps:

1. Click the Upload Files button. Your browser displays the Upload Files To Your Cloud Drive dialog box, shown here.

Upload files to your Cloud Drive	Close

> See upload details

Step 1   Select a Cloud Drive destination folder for your files:

📁 Your Cloud Drive   ▾

Step 2   **Select files to upload...**

File upload size is limited to 2 GB per file

Uploading music? Amazon Cloud Player is the easiest way to upload all your songs.

Storage available: 4.4 GB   Buy additional storage

2. Click the Step 1 button to display the Upload To dialog box, shown here.

Upload To   ☒

⊟ Your Cloud Drive
  ⊟ Documents
    ⊞ Church
    ⊞ Play
    ⊞ Work
  ⊞ Music
  ⊞ Pictures
  ⊞ **Videos**

[Select] [Cancel]

3. Click the folder you want to put the file in. If you want to put the file in a subfolder, click the + button to expand the folder that contains the subfolder.

4. Click the Select button. The Upload To dialog box closes, and the Step 1 button in the Upload Files To Your Cloud Drive dialog box shows the folder you chose.
5. Click the Select Files To Upload button. Your browser displays the Select File(s) To Upload dialog box.
6. Select the file, and then click the OK button. Your browser starts uploading the file.
7. If you want to see the progress of the upload, click the See Upload Details link in the Your Cloud Drive window. The Upload Files To Your Cloud Drive dialog box (shown here) opens and displays the progress. Click the Close button when you want to close the dialog box.

Upload files to your Cloud Drive		Close ☒
⟳ Uploading file: 1 of 1	(73.7 MB / 19 mins remaining)	Stop Upload

All Files (1)

Cloud Drive destination folder: 📁 Videos

| 🎬 boats.avi | 81.8 MB | In progress | 10% | ✕ |

+ Add more files

Storage available: 4.4 GB Buy additional storage

8. When the Upload Complete dialog box appears, click the Close button.

DOUBLE GEEKERY

Refresh the Songs Shown on Cloud Drive

Your Kindle Fire automatically refreshes the list of files on your Cloud Drive every ten minutes. So if you've just uploaded some songs from your computer and they're not yet on Cloud Drive in the Music app, you can simply wait a few minutes, and they'll show up automatically.

If you don't want to wait, you can refresh Cloud Drive manually. Follow these steps:

1. In the Music app, tap the Menu button at the bottom of the screen to display the menu panel.
2. Tap the Settings button to display the Settings screen.
3. Tap the Refresh Cloud Drive button.

Project 4: Convert Songs from Formats Your Kindle Fire Can't Play

If you're buying songs online, or downloading files from artists' websites, you'll normally be able to get MP3 files or AAC files, or sometimes Ogg Vorbis files or WAV files. Your Kindle Fire can play all these types of files, so you can listen to them using the Music app. But other times, you may run into file formats that your Kindle Fire can't play.

This project shows you how to convert songs from the five formats you're most likely to have problems with:

- Protected AAC
- Apple Lossless Encoding
- Free Lossless Audio Codec
- Audio Interchange File Format
- Monkey's Audio

Convert Protected AAC Files

Apple's iTunes Store used to sell songs in the Protected AAC format because the record companies insisted on using digital rights management (DRM) to prevent people from sharing songs freely. Protected AAC files use the .m4p file extension rather than the .m4a file extension, so they're easy to recognize.

Your Kindle Fire can't play Protected AAC files, and iTunes won't let you convert them to other formats. But there are two ways of working around this:

- **Burn the files to CD, and then rip them**   If you have only a few files in the Protected AAC format, add them to a playlist in iTunes, and then burn the playlist to a CD. You can then rip the CD to AAC files or MP3 files. This procedure is awkward, because not only must you use a physical CD, but you must enter the tag information when you rip the CD.

 Burning the protected files to CD and then ripping them is technically illegal in the United States under the Digital Millennium Copyright Act (DMCA).

- **Use iTunes Match to convert the files**   iTunes Match is an iTunes feature for storing your music online in Apple's iCloud service so that you can play it on any of your computers or iOS devices (iPhone, iPad, iPod touch, or Apple TV). iTunes Match costs $24.99 a year at this writing. After you pay for iTunes Match, iTunes scans your library and makes available DRM-free versions of all the matching songs in the iTunes Store; any songs the iTunes Store doesn't have, iTunes uploads to your space on iCloud. You can then download the DRM-free versions of your protected files and use them with your Kindle Fire.

Convert Apple Lossless Encoding Files

Apple Lossless Encoding is a full-quality encoding format. There's no loss of sound quality, but as a result the files are much bigger than MP3 files or AAC files of the same audio. Your Kindle Fire can't play Apple Lossless Encoding files, but it's no big loss—AAC gives you high enough quality at a much smaller file size.

   If you have Apple Lossless Encoding files, you can easily convert them to AAC files or MP3 files using iTunes. Select the files and then choose Advanced | Create AAC Files or Advanced | Create MP3 Files from the menu bar. Only one of these Create Files commands appears; if you want a different file format than the Create Files command shows, change the format as described in the section "Set Up iTunes for Ripping Your CDs" in Project 1.

DOUBLE GEEKERY

Identify Apple Lossless Encoding Files Masquerading as AAC Files

The Apple Lossless Encoding format uses the same file extension, .m4a, as the AAC file format—so if you just look at the filename, you can't tell which format the file uses.

   If you find an .m4a file won't play on your Kindle Fire, chances are it's Apple Lossless Encoding rather than AAC. To check, follow these steps in iTunes:

1. Right-click (or CTRL-click on the Mac) the song to display the context menu, and then click Get Info to display the Item Information dialog box for the song. You can also click the song and then press CTRL-I on Windows or ⌘-I on the Mac.
2. Click the Summary tab if it's not already displayed. The next illustration shows the top part of the Summary tab.

Bobby Moore's Wine	

Bobby Moore's Wine (3:22)
Chameleons
Free Trade Hall Rehearsal

Kind: Apple Lossless audio file   **Channels:** Stereo
Size: 22.4 MB   **Sample Size:** 16 bit
Bit Rate: 922 kbps   **Encoded with:** iTunes 8.1.1
Sample Rate: 44.100 kHz
Date Modified: 5/8/2009 6:32 AM
Plays: 25
Last Played: 11/10/2011 9:34 AM

3. Look at the Kind readout to see whether it says AAC Audio File or Apple Lossless Audio File.
4. Click the OK button to close the Item Information dialog box.

Convert Free Lossless Audio Codec Files

Free Lossless Audio Codec (FLAC) is an open-source full-quality encoding format—the open-source equivalent of Apple Lossless Encoding, as it were. Your Kindle Fire can't play FLAC files, but they're so big that you probably wouldn't want to load them on the Kindle Fire anyway.

FLAC files use the .flac file extension, so they're easy to identify. If you have FLAC files, use a utility such as Total Audio Converter (Windows; www.coolutils.com) or X Lossless Decoder (Mac; search for it) to create AAC files or MP3 files.

Convert Audio Interchange File Format Files

Audio Interchange File Format (AIFF) is an uncompressed audio format similar to WAV but mostly used on the Mac. Your Kindle Fire can't play AIFF files, so if you have any, use iTunes to create AAC files or MP3 files from them. AIFF files use the .aiff file extension.

Select the files, open the Advanced menu, and then click the Create Files command (for example, Create AAC Files). To use a different file format than the Create Files command shows, change the format as described in the section "Set Up iTunes for Ripping Your CDs" in Project 1.

Convert Monkey's Audio Files

Monkey's Audio is a lossless encoding format that's popular among serious audiophiles. It's not widely used, partly because encoding and decoding requires a lot of processing power—more than most portable players have. Monkey's Audio uses the .ape file extension, which is memorable.

iTunes can't handle Monkey's Audio files, so if you get Monkey's Audio files, you need to use other applications to convert them.

Convert Monkey's Audio Files to MP3 or AAC on Windows

On Windows, your best bet is to decompress the Monkey's Audio file to a WAV file, and then use iTunes to convert the WAV file to whichever format you want.

First, download the latest version of Monkey's Audio from the Monkey's Audio website (www .monkeysaudio.com) and install it. On the final screen of the install wizard, leave the check box selected to launch Monkey's Audio for you, and click the Finish button. Monkey's Audio then opens, as shown here.

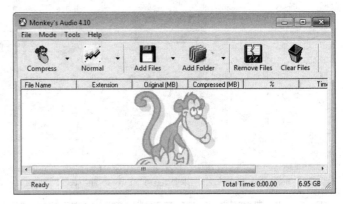

With Monkey's Audio open, you can convert Monkey's Audio files to WAV files like this:

1. Choose Mode | Decompress to switch to Decompress mode.
2. Choose Tools | Options to display the Options dialog box, and then click the Output category in the left pane to display the Output controls (shown here).

3. In the Output Location area, select the Output To Specified Directory option button.
4. Click the Browse button to display the Browse For Folder dialog box, click the folder you want to put the converted files in, and then click the OK button.
5. Click the OK button to close the Options dialog box.
6. Click the Add Files button to display the Open dialog box.
7. Select the file or files you want to decompress, and then click the Open button. The Open dialog box closes, and the file or files appear in the Monkey's Audio window, as shown here.

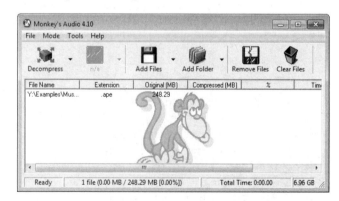

8. Click the Decompress button on the toolbar. Monkey's Audio decompresses the file to a WAV file.

You can now import the WAV file into iTunes and then convert it to AAC or MP3, as discussed earlier in this chapter.

Convert Monkey's Audio Files to MP3 or AAC on the Mac

To convert Monkey's Audio files to MP3 or AAC on the Mac, use Max. Follow these steps to get Max and set it up:

1. Download the latest version of Max from the sbooth.org website (http://sbooth .org/Max/).
2. If your Mac doesn't automatically unzip the distribution file and display its contents in a Finder window, click the Downloads icon on the Dock, and then click the Max distribution file to unzip it and display its contents.
3. In the Max folder, double-click the Max application file to launch Max. You'll see the Max menus on the menu bar, but no Max window at first.

 You can either run Max from its folder or drag the application to your Applications folder so that you can run it from Launchpad. While trying out Max, you may prefer to run the app from its folder.

4. Choose Max | Preferences to display the Preferences window.
5. Click the Formats tab to display the Formats pane (shown in Figure 1-24 with MP3 already configured as an output format).
6. If you want to convert your Monkey's Audio files to MP3 files, follow these substeps:

   a. Double-click MP3 in the Available Output Formats list box to display the MP3 dialog box (see Figure 1-25).

   b. In the Encoder Quality pop-up menu, choose the quality you want: Best (for highest quality), Transparent (for CD quality), Portable (for acceptable quality and small files), or Custom. If you choose Custom, choose settings for all the other controls in the MP3 dialog box.

   c. Click the OK button. The MP3 dialog box closes, and MP3 appears in the Configured Output Formats list box.

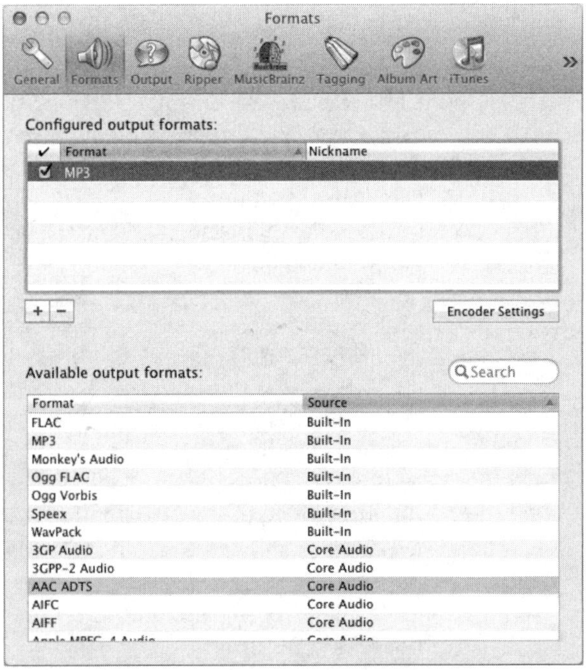

FIGURE 1-24 Set up your preferred output format in the Formats pane of Max's Preferences window.

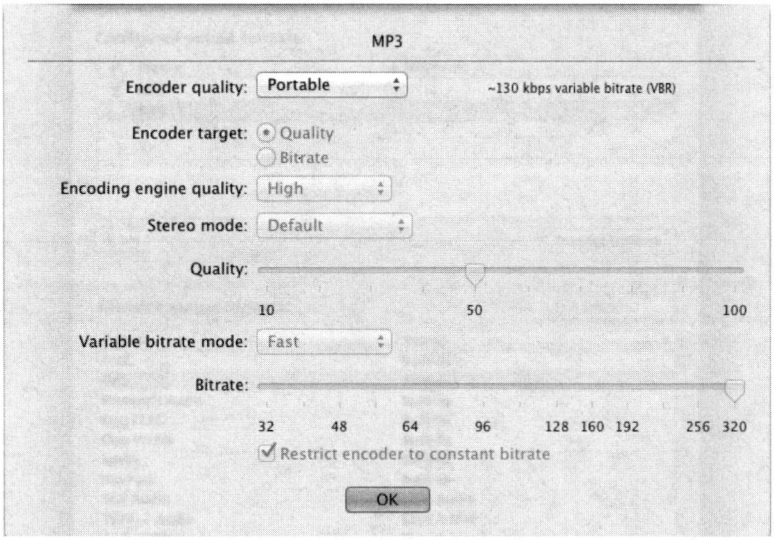

FIGURE 1-25 In the MP3 dialog box, choose the quality of MP3 files
you want to create. Choose Portable if you want to pack as much music on
your Kindle Fire as possible.

7. If you want to convert your Monkey's Audio files to AAC files, follow these
    substeps:
    a. Double-click Apple MPEG-4 Audio in the Available Output Formats list box
        to display the Apple MPEG-4 Audio (AAC) dialog box (see Figure 1-26).
    b. In the Available Data Formats list box, click AAC.
    c. In the Extension pop-up menu, make sure AAC is selected.
    d. In the Quality pop-up menu, select the quality you want—for example, High.
    e. In the Bitrate pop-up menu, select the bitrate you want—for example, 128 Kbps.
    f. Select the Use VBR check box if you want to use variable-bitrate encoding.
        This normally gives better results.
    g. Click the OK button. The Apple MPEG-4 Audio (AAC) dialog box closes, and
        Apple MPEG-4 Audio appears in the Configured Output Formats list box.
8. In the Configured Output Formats list box, select the check box for the format
    you want to use—either Apple MPEG-4 Audio or MP3.
9. Click the Output tab to display its controls (see Figure 1-27).
10. Open the Output Files pop-up menu and choose where to store the converted
    files. You can choose Same As Source File to have Max use the folder the original
    files are in, or click the Choose Directory button and use the resulting dialog box
    to choose the folder.
11. Click the close button (the red button at the left end of the title bar) to close the
    Preferences window.

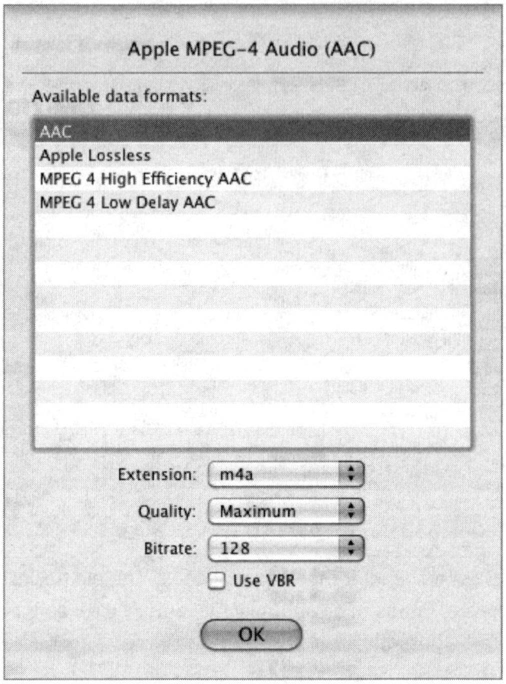

FIGURE 1-26 In the Apple MPEG-4 Audio (AAC) dialog box, select the AAC item, and then choose the extension, quality, and bitrate. You can also choose whether to use variable-bitrate encoding.

After setting up Max, you can quickly convert files with it. Follow these steps:

1. Choose File | Open to display the Open dialog box.
2. Select the file or files you want to convert to your preferred output format.
3. Click the Open button. Max displays a window with details of the tag information and tracks (see Figure 1-28).
4. Correct or complete the tag information as needed.
5. In the list of tracks, select the check box for each song you want to convert. Normally, you'll want to click the Select All button to select all the check boxes at once.

6. Click the Convert button. Max converts the songs to the format you chose, displaying the Encoder dialog box (shown here) as it does so.

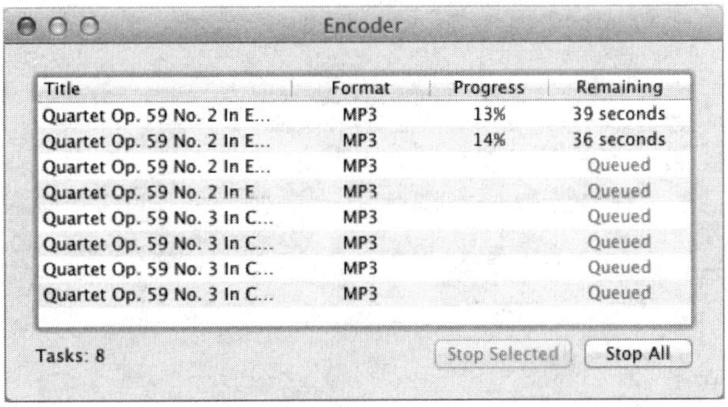

When Max finishes converting the files, you can copy them directly to your Kindle Fire or add them to your iTunes Library and then put them on the Kindle Fire from there.

FIGURE 1-27   Specify your output folder in the Outputs pane of Max's Preferences window.

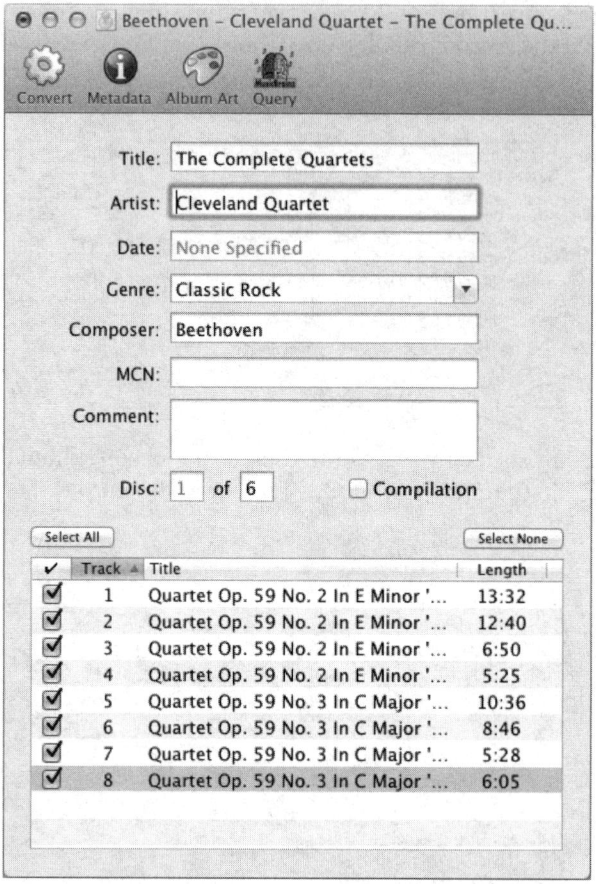

FIGURE 1-28   In the Max window, correct or complete the tag information, and choose which songs to convert.

Project 5: Listen to the Radio on Your Kindle Fire

In the previous projects, you've seen how to make the most of your Kindle Fire's limited storage by cramming compressed music files onto it and by playing other files from Cloud Drive.

Having your own music available wherever you are is great, but you can also enjoy music on the Kindle Fire by listening to the radio. This gives you a wide range of different music (and other audio—for example, talk radio and sports radio) without consuming any of the Kindle Fire's precious storage space.

The Kindle Fire works pretty well as a radio because it's small enough to carry with you most anywhere, and its speakers produce enough volume for comfortable listening, especially for talk radio rather than music. (For music, you may want to connect your Kindle Fire to speakers, as discussed in Project 7, later in this chapter.)

To listen to the radio on your Kindle Fire, first you need to get a suitable radio app, because the Kindle Fire doesn't include one. Getting the app takes a couple of minutes. You can then tune in and start listening.

DOUBLE GEEKERY

Create Your Own Personalized Radio Stations with Pandora

Although your Kindle Fire doesn't include a radio app, its Apps screen does include a link to the Pandora app. Pandora is a special Internet radio app that lets you create your own personalized radio stations—ones that play only the music you like and that adapt to learn your musical taste.

To get started with Pandora, follow these steps:

1. Tap the Home button to display the Home screen.
2. Tap the Apps button to display the Apps screen.
3. Tap the Device tab at the top if it's not already active.
4. Tap the Pandora icon. Your Kindle Fire downloads the Pandora app from the Appstore, installs it, and launches it.
5. Tap the button for creating a new account, and then follow through the setup process. On the Create A Free Account screen, clear the Send Me Personalized Recommendations And Tips check box (which is selected by default) unless you actually want to receive advertising messages from Pandora.
6. On the screen shown here, set up your first Pandora station:
   a. Tap the Artist tab or the Genre tab, as appropriate.

```
Guy's kindle  3                    5:23                    ⚙ 📶 🔋
                          [ Artist ][ Genre ]

   [ artist, song, or composer                                    ]

   [ Search ]
```

   b. Type your search term into the text box. For example, type an artist's name.
   c. Tap the Search button. Pandora displays matches.
   d. Tap the match you want. Pandora starts playing a track by the artist.

7. Use the controls at the bottom of the Pandora window (shown here) to control playback and rate the music:

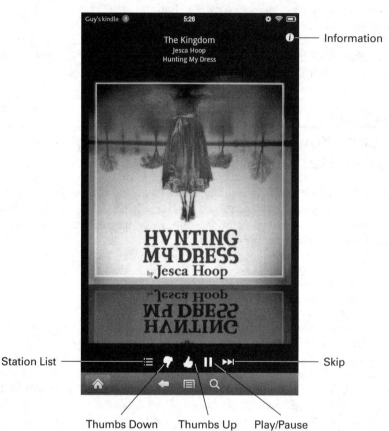

Station List ———

Information

Skip

Thumbs Down   Thumbs Up   Play/Pause

- **Information**   Tap this button to display information about the artist and the song. Tap the cover picture button (which replaces the i button) to return to the cover picture.
- **Station List**   Tap this button to display your list of stations. From here, you can pick an existing station or start creating a new one.
- **Thumbs Down**   Tap this button to mark the current song as one you don't like. Pandora automatically skips to the next song.
- **Thumbs Up**   Tap this button to mark the current song as one you like. Pandora keeps playing the song.
- **Play/Pause**   Tap this button to start, pause, and resume playback.
- **Skip**   Tap this button to skip to the next song.

8. When you are ready to stop using Pandora, tap the Menu button to display the menu panel (shown here), and then tap the Quit button.

Because of some complicated legal issues involving copyright, Pandora is available only in the United States at this writing. Pandora is hoping to add other countries soon. If, when you launch Pandora, you see a message saying Pandora isn't available in your country, you're out of luck. The only way around this is to use a virtual private network (VPN) service to give your Kindle Fire an IP address within the United States. See the section "Access Cloud Player from Outside the United States," earlier in this chapter, for a couple of VPN services to try.

Get a Radio App for the Kindle Fire

If you want to listen to the radio on your Kindle Fire, you need to get a radio app. You can get a good variety of radio apps for free from the Amazon Appstore. This section shows screens using the popular app called TuneIn Radio, but you may prefer another app.

To get a radio app, follow these steps:

1. Tap the Home button to display the Home screen.
2. Tap the Apps button to display the Apps screen.
3. Tap the Store button in the upper-right corner to display the Store screen (see Figure 1-29).
4. Tap in the Search In Appstore box at the top, type **radio** on the virtual keyboard, and then tap the Search button. Your Kindle Fire displays a list of matches, as shown in Figure 1-30.
5. If the app is free, tap the Free button and then tap the Get App button. If the app has a price, tap the price button, and then tap the Buy App button that replaces it.
6. Wait while your Kindle Fire downloads and installs the app, and then tap the Open button on the app's information screen. The app then opens.

Listen to Your Radio App

After installing and opening your radio app, you can quickly start listening to the stations you like. For example, in TuneIn, you can browse stations by using the Browse screen (see Figure 1-31).

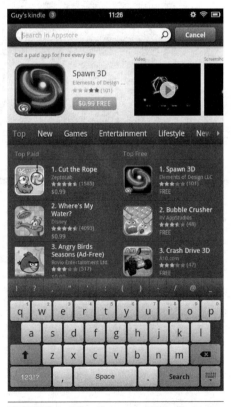

FIGURE 1-29   You can quickly find a radio app for your Kindle Fire by searching the Appstore.

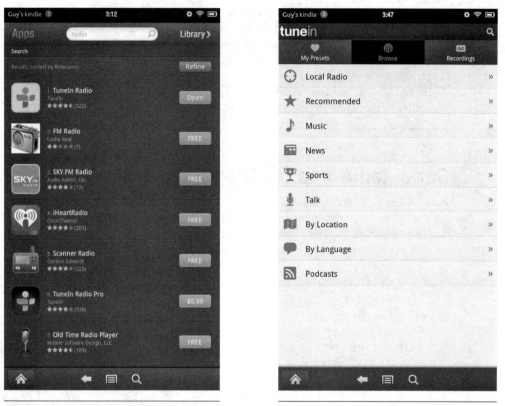

FIGURE 1-30 The Appstore provides a healthy selection of both free and pay radio apps.

FIGURE 1-31 Use the Browse screen in the TuneIn radio app to find the types of stations you prefer.

When you find a station you want to listen to, tap it. The station starts playing, and you can control playback by using the controls. For example, in TuneIn, you can pause or stop the audio, or start recording it, by tapping the buttons in the lower-right corner (see Figure 1-32). Tap the Preset button (the heart button in the upper-right corner) to create a preset for the station so that you can quickly access it from the Presets tab in the future.

FIGURE 1-32   Tap the buttons in the lower-right corner of the TuneIn screen to control playback or record the radio show.

Project 6: Enjoy Podcasts on Your Kindle Fire

A *podcast* is a downloadable show that you can play on your Kindle Fire or on your computer. Some podcasts are downloadable versions of professional broadcast radio shows, while others are put together by enthusiasts.

 I can see you're about to ask what the difference is between an Internet radio show and a podcast. The answer is: It depends. Radio shows by conventional radio stations are broadcast in real time on the Internet as well as on radio waves, but most shows are also available for playback later on demand. Podcasts are typically not broadcast in real time but are available for streaming playback or for download whenever it suits you.

In this section, we'll look at how to enjoy podcasts on your Kindle Fire. You can either download podcasts to your Kindle Fire so that you can play them back later or stream the podcasts across the Internet via a Wi-Fi connection.

Get an App That Can Play Podcasts

Your Kindle Fire doesn't include an app that can play podcasts from the Internet, so normally your first move is to install an app to play podcasts. You can find a good selection of podcast apps on the Amazon Appstore by searching using the term "podcast."

In this section, we'll look at what's arguably the best podcast app: BeyondPod Podcast Manager. BeyondPod Podcast Manager comes in a Lite version that's free and a Pro version that costs $6.99. The Lite version gives you full Pro functionality for the first seven days to tempt you to pay for the Pro version.

To get BeyondPod Podcast Manager, go to the Appstore and search for it. See the section "Get a Radio App for the Kindle Fire" in the previous project for instructions on accessing the Appstore, finding apps, and installing them.

Run BeyondPod Podcast Manager

After installing BeyondPod Podcast Manager, tap the Open button on the app's information screen to launch it. BeyondPod Podcast Manager displays the BeyondPod Quick Start screen. Read the information this screen provides, and then tap the Back button.

You then see the Feeds View screen, which shows the various feeds you've subscribed to (see Figure 1-33).

At this point, you haven't actually subscribed to any feeds, but BeyondPod Podcast Manager provides you with a selection of default feeds to get you going.

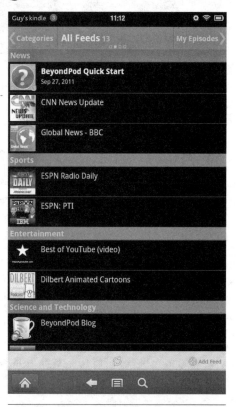

Add and Remove Feeds

To add a feed, tap the Add Feed button at the lower-right corner of the Feeds View screen. On the Add Feed screen (see Figure 1-34) that BeyondPod Podcast Manager displays, you can add a feed in any of these ways:

- Type the feed's name or URL in the Search By Name Or Topic box at the top of the screen. Tap the Search button (the magnifying-glass icon) to display matching results. You can then tap the Preview button to preview a podcast or the Subscribe button to subscribe to it.

FIGURE 1-33 From the Feeds View screen in BeyondPod Podcast Manager, you can browse the default selection of podcasts and feeds and add your preferred items.

Guy's kindle ③   12:58   ⚙ 📶 🔋

Add Feed

[Search by name or topic   ] [🔍]

*If you know the feed URL, you can type or paste it in the search box above (start with **http://**)*

Feed search tips

Browse Popular Feeds

🌐 Popular Feeds and Podcasts

🌐 NPR Podcast Directory

🌐 Wurl.com Video Guide

More Directories...

Import Feeds

Import from Google Reader   ❯

Import from SD card Folder   ❯

Import from OPML file   ❯

FIGURE 1-34   From the Add Feed screen in BeyondPod Podcast Manager, you can add a feed by typing its name or URL, by browsing for the feed, or by importing a list of feeds.

- In the Browse Popular Feeds area, tap the button for the directory you want to browse. You can then navigate through the feeds by categories—for example, News, Business And Finance, or Technology—to find a feed that interests you. When you tap the feed, BeyondPod Podcast Manager displays the Add To Category dialog box (shown here). Tap the category to which you want to add the feed.

- In the Import Feeds area, tap the appropriate button—Import From Google Reader, Import From SD Card Folder, or Import From OPML File—and then follow the instructions to import a list of feeds.

OPML is the abbreviation for Outline Processor Markup Language, an XML format used for creating outlines or collapsible lists—in this case, a list of feeds.

To remove a feed from the Feeds View screen, tap and hold the feed's button until the dialog box shown here appears, and then tap the Delete Feed button.

Update Feed
Delete Feed
Edit Feed
Share Link

View the Feed Content Available

To see the episodes available in a feed, tap its button on the Feeds View screen. BeyondPod Podcast Manager displays the Feed Content View screen (see Figure 1-35), which shows the episodes in a list.

FIGURE 1-35 The Feed Content View screen in BeyondPod Podcast Manager shows a list of the episodes in a particular feed.

FIGURE 1-36   Use the controls on the Player View screen to control playback.

From here, you can tap the Download button to download a podcast episode to your Kindle Fire so that you can play it later, or you can simply tap the Stream button to start playing the episode.

Play a Podcast Episode

When you start playing a podcast episode, BeyondPod Podcast Manager displays the Player View screen (shown in Figure 1-36 in landscape view). You can then use the buttons to control playback.

Project 7: Use Your Kindle Fire as Your Home Stereo

Your Kindle Fire is great for music on the go, but you can use it as your home stereo as well. In this project, we'll look at your options: using a Kindle Fire speaker dock or a pair of powered speakers, and connecting your Kindle Fire to your existing stereo via either a wired connection or via a radio transmitter.

Use a Kindle Fire Speaker Dock or a Pair of Powered Speakers

The simplest way to get a decent volume of sound from your Kindle Fire is to connect it to a pair of powered speakers—speakers that include their own amplifier. You can find a huge range of powered speakers in both online and real-world stores, with prices running from a few dollars to the high hundreds of dollars.

If you've already got a pair of powered speakers, all you need do is plug their miniplug connector into your Kindle Fire's headphone socket, and you'll be in business.

 If your powered speakers have a volume control, set your Kindle Fire's volume to a moderate volume—around 50 percent or 60 percent—and then use the speakers' volume control to adjust the overall volume. If your powered speakers don't have a volume control, you can either add one built into a cable or simply control the volume from your Kindle Fire.

If you don't have a pair of powered speakers, or you're looking for a more elegant solution, you can buy powered speakers designed especially for the Kindle Fire. Such speakers typically include a docking holder of some kind and a power feed for the Kindle Fire, so you can charge it via its USB port while you play music (or videos).

At this writing, there are relatively few speaker systems designed for the Kindle Fire—but the market is growing. Here are a couple of examples:

- **MobiDock** The MobiDock from iLuv ($99.99; www.i-luv .com and various online stores, including Amazon) is an audio dock for the Kindle Fire and for smart phones that have micro-USB connections. Figure 1-37 shows the MobiDock.
- **FireStation** The FireStation from ReaderDock ($59.99; www.readerdock.com and other online stores, including Amazon) is an audio dock designed specifically for the Kindle Fire. The FireStation (see Figure 1-38) has built-in speakers, provides a micro-USB power feed, and works in both portrait and landscape orientations.

FIGURE 1-37 The MobiDock has adjustable arms to hold either the Kindle Fire or a smart phone, and provides power via a micro-USB connector. (Image courtesy of iLuv Creative Technology.)

FIGURE 1-38   The FireStation can hold the Kindle Fire in either portrait or landscape orientation, provides speakers, and delivers a power feed. (Image courtesy of ReaderDock.)

 When choosing a Kindle Fire speaker system, look for one that enables you to dock the Kindle Fire in both landscape mode and portrait mode so that you can use the speakers for watching movies comfortably as well as for playing back music.

Use a Wired Connection to Your Existing Speakers or Stereo

The most direct way to connect your Kindle Fire to a stereo system is with a cable. For a typical receiver, you'll need a cable that has a miniplug at the end you'll connect to your Kindle Fire and two RCA plugs at the other end, as shown in Figure 1-8, earlier in this chapter. Figure 1-39 shows an example of a Kindle Fire connected to a stereo via the amplifier.

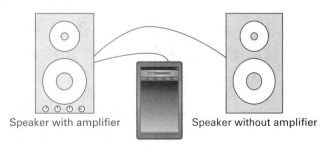

Speaker with amplifier   Speaker without amplifier

FIGURE 1-39   A miniplug-to-RCA-plugs cable is the most direct way of connecting your Kindle Fire to your stereo system.

Some receivers and boom boxes use a single stereo miniplug input rather than two RCA ports. To connect your Kindle Fire to such devices, you'll need a stereo miniplug-to-miniplug cable. Make sure the cable is stereo, because mono miniplug-to-miniplug cables are common. A stereo cable has two bands around the miniplug (as on most headphones), whereas a mono cable has only one band.

To make sure you get good quality audio out of the Kindle Fire, get a high-quality cable rather than a bargain-basement one. A low-quality cable can degrade the audio a surprising amount.

Connect your Kindle Fire to your receiver as follows:

1. Connect the miniplug to your Kindle Fire's headphone port.
2. Tap the status bar to display the Settings bar, tap the Volume button, and then drag the Volume slider all the way to the left to turn down the volume.
3. Turn down the volume on the amplifier as well.
4. Connect the RCA plugs to the left and right ports of one of the inputs on your amplifier or boom box—for example, the AUX input or the Cassette input (if you're not using a cassette deck).

 Don't connect the Kindle Fire to the Phono input on your amplifier. The Phono input is built with a higher sensitivity to make up for the weak output of a record player. Putting a full-strength signal into the Phono input will probably blow it.

5. Start the music playing.
6. Turn up the volume a little on your Kindle Fire.
7. Turn up the volume on the receiver so that you can hear the music.
8. Increase the volume on the two controls in tandem until you reach a satisfactory sound level.

 Too low a level of output from your Kindle Fire may produce noise as your amplifier boosts the signal. Too high a level of output from your Kindle Fire may cause distortion.

Use a Radio Transmitter Between Your Kindle Fire and a Stereo

If you don't want to connect your Kindle Fire directly to your stereo, you can use a radio transmitter to send the audio from your Kindle Fire to the radio on your stereo. This device plugs into the Kindle Fire's headphone port and broadcasts a signal on an FM frequency to which you then tune your radio to play the music. Most radio transmitters offer a choice of frequencies to allow you easy access to both your Kindle Fire and your favorite radio stations.

Radio transmitters can deliver reasonable audio quality. If possible, try before you buy by asking for a demonstration in the store (take a portable radio with you, if necessary).

Find a Suitable Frequency for a Radio Transmitter

In most areas, the airwaves are busy these days—so to get good reception on your radio from your Kindle Fire's radio transmitter, you need to pick a suitable frequency. To do so, follow these steps:

1. With the Kindle Fire's radio transmitter turned off, turn on your radio.
2. Tune the radio to a frequency on which you get only static and for which the frequencies one step up and one step down give only static as well. For example, if you're thinking of using the 91.3 frequency, make sure that 91.1 and 91.5 give only static as well.
3. Tune the radio transmitter to the frequency you've chosen, and see if it works. If not, identify and test another frequency.

This method may sound obvious, but what many people do is pick a frequency on the radio transmitter, tune the radio to it—and then are disappointed by the results.

Radio transmitters are relatively inexpensive (usually between $15 and $50, but it's easy enough to pay more if you want to) and easy to use (but see the nearby sidebar for instructions on tuning).

The sound you get from this arrangement typically will be lower in quality than the sound from a wired connection, but it should be at least as good as listening to a conventional radio station in stereo. If that's good enough for you, a radio transmitter can be a neat solution to playing music from your Kindle Fire throughout your house. You can also use them to play music through multiple radios (or sets of speakers connected to radios) at once, giving yourself music throughout your dwelling without complex and expensive rewiring.

Three Other Uses for a Radio Transmitter

A radio transmitter is great for playing music from your Kindle Fire through your stereo. But you'll probably also want to use your radio transmitter for other purposes too:

- **Car** As you'll see in the next project, radio transmitters are good for use in the car as well—especially if your car doesn't have any other way to connect your Kindle Fire as an audio source.
- **PC or Mac** You can use a radio transmitter to broadcast audio from your PC or Mac to a radio. This is a handy way of getting streaming radio from the Internet to play on a conventional radio when you're using a computer rather than your Kindle Fire.
- **Travel** When you travel, take your radio transmitter along so that you can play music through a hotel radio.

Project 8: Use Your Kindle Fire as Your Car Stereo

If you've loaded your Kindle Fire with your favorite music and carry the Kindle Fire everywhere with you, you probably want to use it to play music through your car stereo. In this project, you learn the different ways in which you can play music from the Kindle Fire through your car stereo and choose the way that suits you best. We also consider where to put the Kindle Fire while it's playing—and how to keep it there.

Assess Your Options for Playing Music Through Your Car Stereo

Depending on what type of car stereo you have, you'll probably have three main options for playing music from your Kindle Fire through it:

- Wire the Kindle Fire directly to the car stereo and use it as an auxiliary input device.
- Use a radio-frequency device to play the Kindle Fire's output through the car's radio.
- Use a cassette adapter to connect the Kindle Fire to the car's cassette player.

Each of these methods has its pros and cons. The following sections tell you what you need to know to choose the best option for your car stereo.

Wire Your Kindle Fire Directly to a Car Stereo

If you want the best audio quality possible, see if you can connect your Kindle Fire directly to your car stereo. How easily you can do this depends on how the stereo is designed:

- If your car stereo has a miniplug input built in, get a miniplug-to-miniplug cable to connect the Kindle Fire's headphone port to the miniplug input.
- If your stereo is built to take multiple inputs—for example, a CD player (or changer) and an auxiliary input—you may be able to simply run a wire from unused existing connectors. Then all you need to do is plug your Kindle Fire into the other end and press the correct buttons to get the music going.
- If no unused connectors are available, you or your local friendly electronics technician may need to get busy with a soldering iron.

Use a Radio Transmitter with Your Kindle Fire

If the car stereo doesn't have a miniplug input or spare connection, your easiest option for playing music from your Kindle Fire may be to get a radio transmitter. Look for a model that can take power from your car's accessory socket so you're not continually running through the device's batteries.

See the previous project for a discussion of the pros and cons of a radio transmitter and instructions on how to find a suitable frequency. You may need to switch frequencies as you drive through areas with radio stations broadcasting on different frequencies, so pick several frequencies that seem uninhabited and set one of your car radio's presets to each.

Use a Cassette Adapter with Your Kindle Fire

If the car stereo has a cassette player, you can use a cassette adapter to play audio from your Kindle Fire through the cassette deck. You can buy such adapters for between $10 and $20 from most electronics stores either online or offline.

The adapter is shaped like a cassette and uses a playback head to input analog audio via the head that normally reads the tape as it passes. A wire runs from the adapter to your input source—in this case, the Kindle Fire.

A cassette adapter can be an easy and inexpensive solution, but it's far from perfect. The main problem is that the audio quality tends to be poor, because the means of transferring the audio to the cassette player's mechanism is less than optimal. But if your car is noisy, you may find that road noise obscures most of the defects in audio quality.

If the cassette player's playback head is dirty from playing cassettes, audio quality will be that much worse. To keep the audio quality as high as possible, clean the cassette player regularly using a cleaning cassette.

 If you use a cassette adapter in an extreme climate, try to make sure you don't bake it or freeze it by leaving it in the car.

Mount Your Kindle Fire in Your Car

At a pinch, you can place your Kindle Fire on the passenger seat and let it slide from side to side as you weave through traffic or chicanes. But normally you'll want to mount your Kindle Fire securely in your car within reach. That way, you can keep it connected to your car stereo via a cable (if it needs one) and connected to a power feed (likewise).

You can get various devices to mount your Kindle Fire in the front of your car and keep it within reach:

- **Windshield mount**   Some people swear by these mounts, which use a sucker cup to secure the Kindle Fire or other device to the windshield via a flexible neck. But most people find they work better with smaller devices, such as iPhones or other mobile phones.
- **Air vent mount**   Mounting the Kindle Fire on an air vent is usually more secure than using the windshield, and in most cars it puts the Kindle Fire closer to hand. But because most air vent mounts are designed for smart phones and other small devices, you need to make sure that the mount you get is large and sturdy enough for the Kindle Fire.

- **Cup holder mount** If your car has a cup holder to spare, a cup holder mount can be a great solution, because it's plenty strong enough and is usually in a safe position. Figure 1-40 shows the Kindle Fire fixed on a cup holder mount and wired to the stereo.
- **Dashboard mount** If none of the other mount types is viable, consider a dashboard mount. Most of these use suction cups, which may struggle to get enough grip on the dashboard to hold the Kindle Fire securely. You also need to position the mount where neither you nor the airbag will collide with the Kindle Fire in a crash.

Power Your Kindle Fire in Your Car

If you use your Kindle Fire extensively in your car, you'll probably want to charge it there as well. You can easily do so by getting a car charger that connects to your car's accessory socket.

You can get chargers specifically designed for the Kindle Fire, but a generic charger will do fine as well. The Kindle Fire needs a current of at least 1800 milliamps (1.8 amps) to charge it, so make sure the charger you get delivers this current or more. For best results, get a charger that gives 2100 milliamps (2.1 amps).

As well as the charger unit, you'll need a USB cable from 2.0A Male to

FIGURE 1-40 Mounting the Kindle Fire in a cup holder is often both easy and secure.

Micro B to convey the power to the Kindle Fire. Make the connections, and you'll be in business.

2 Video Geekery

With its bright, clear screen, your Kindle Fire is great for playing back video. You can play back videos you buy or stream from Amazon using the built-in Video app, or load your own videos from your computer or other sources and play them back using the built-in Gallery app.

Playing video is straightforward, so we'll skip lightly over that and start by adding TV shows to your Kindle Fire. We'll then look at how to stream movies from your computer to your Kindle Fire and how to play a movie from your Kindle Fire on your computer.

After that, I'll show you how to put your videos and DVDs on your Kindle Fire. Finally, we'll look at how to use your Kindle Fire for in-car entertainment on those long journeys.

Project 9: Watch TV Shows on Your Kindle Fire

If you open the Videos app, tap the Store tab, and then tap the TV Shows bar, you'll find various TV shows: *Glee, Grey's Anatomy, The Office,* and so on. These are shows you can buy by the episode, download to your Kindle Fire, and keep.

But if you want to catch up with current TV, you need to turn to a streaming solution. At this writing, the two leading providers of streaming TV programs are Hulu and Netflix.

Watch TV Shows on Hulu Plus

Hulu's basic service, called simply Hulu, provides free, ad-supported streaming of TV shows and movies—but you can't use it on your Kindle Fire. Instead, you need to pay for a subscription to Hulu Plus.

Set Up a Hulu Plus Account

At this writing, Hulu Plus costs $7.99 per month. Hulu provides a one-week free trial so that you can see whether the service suits you. As you'd expect, you need to provide a valid credit card for the trial—and you need to cancel within those seven days if you don't want to start paying.

At this writing, Hulu is available only in the United States and Japan. If you're located elsewhere, you'll need to use a web proxy or a VPN connection to make Hulu think your computer is in one of those countries. See the Double Geekery sidebar "Access Cloud Player from Outside the United States" in Project 3 (Chapter 1) for some suggestions.

Go to the Hulu website (www.hulu.com), either on your computer or on your Kindle Fire, and then click any of the Hulu Plus links. Click the Sign Up link, type your e-mail address, and fill in the membership form.

Install the Hulu Plus App on Your Kindle Fire

Once you've set up your Hulu Plus account, install the Hulu Plus app on your Kindle Fire. Tap the Apps button on the Home screen, tap the Store button on the Apps screen, and then type **hulu** into the Search box. Tap the Free button on the Hulu Plus (Kindle Fire Edition) search result, and then tap the Get App button.

When your Kindle Fire finishes downloading and installing Hulu Plus, tap the Open button to open the app. On the opening screen (see Figure 2-1), tap the Log In button.

FIGURE 2-1 On the opening screen of the Hulu Plus app, tap the Log In button.

On the login screen (shown here), type your e-mail address and password, and then tap the Log In button.

Find and Watch TV Shows on Hulu Plus

After Hulu Plus logs you in, you see the screen shown next. Tap the TV button, and you can start browsing the TV listings (see Figure 2-2). When you find a show you want to watch, tap it to start it streaming.

Watch TV Shows on Netflix

Like Hulu Plus, Netflix offers streaming movies and TV shows. To use Netflix, you sign up for a Netflix subscription using a web browser, and then install the Netflix app on your Kindle Fire.

Set Up a Netflix Account

Also like Hulu Plus, a Netflix account costs $7.99 per month. Netflix offers a one-month free trial to get you started. You'll need a credit card to sign up, and you'll need to remember to cancel the trial if you don't want to be charged.

To set up a Netflix account, steer your web browser to the Netflix website, www .netflix.com

Install the Netflix App on Your Kindle Fire

After creating your Netflix account, download the Netflix app from the Appstore and install it on your Kindle Fire. Tap the Apps button on the Home screen, tap the Store button on the Apps screen, and then type **netflix** into the Search box. Tap the Netflix For Kindle Fire suggestion, tap the Free button on the Netflix search result, and then tap the Get App button.

FIGURE 2-2 When you find a show you want to watch, tap its button to start it streaming.

When your Kindle Fire finishes downloading and installing Netflix, tap the Open button to open the app. On the sign-in screen (shown here), type your e-mail address and password, and then tap the Sign In button.

Find and Watch TV Shows on Netflix

After signing in, you see the main Netflix screen (shown here), which shows categories such as Critically-Acclaimed Movies, Inspiring Movies, Mind-Bending Movies, and Cerebral Movies. If you're after movies, you can find them by looking through the categories, tapping the Browse button and using the Browse panel, or tapping the Search box in the upper-right corner and typing your search terms.

To find TV shows, tap the Browse button to display the Browse panel (shown here), and then tap the TV Shows button.

You'll then see the TV Shows screen (shown here), which you can browse by category or by searching.

Tap a show to see its information window (shown here), and then tap the Play button for the show you want to play. You can then control playback using the pop-up onscreen controls (see Figure 2-3).

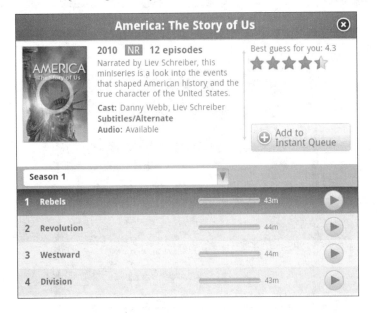

FIGURE 2-3   To control playback in Netflix, tap the screen and then use the pop-up controls.

Project 10: Stream Movies from Your Computer to Your Kindle Fire

Streaming movies (and TV shows) from online services such as Amazon Prime, Hulu Plus, or Netflix is great, but other times you'll want to play back on your Kindle Fire the videos and movies you have on your computer. One way to do so is by loading the files on your Kindle Fire, as described in the next section, first converting them to a Kindle Fire–friendly format if necessary. But you can also stream your movies and videos from your computer to your Kindle Fire.

In this project, I'll show you how to stream movies using Splashtop. This app costs $4.99 at this writing, but it's easy to use and very effective.

Here's what you'll need to do this:

- Install the Splashtop Streamer application on your PC or Mac.
- Install the Splashtop app on your Kindle Fire.
- Set the video streaming on your PC or Mac.
- Set the stream playing on your Kindle Fire.

Install Splashtop Streamer on Your PC or Mac

First, install the Splashtop Streamer application on your PC or Mac. Follow these steps:

1. Open your web browser and go to the Splashtop Streamer download page on the Splashtop website (www.splashtop.com/streamer).
2. Click the Download Splashtop Streamer button. This button shows the operating system the site has detected you're using, so it's called either Download Splashtop Streamer For Windows or Download Splashtop Streamer For Mac.
3. When the download finishes, install the application as usual:
   - **Windows**   Click the Run button to launch the Splashtop Streamer installation, and then wait while the InstallShield Wizard installs Splashtop Streamer. On the InstallShield Wizard Complete screen, click the Finish button. Splashtop Streamer then runs. Click the Go button on the introductory screen, and then click the Accept button on the License Agreement screen.
   - **Mac**   If OS X doesn't automatically open a Finder window showing the contents of the Splashtop Streamer disk image, click the Downloads icon on the Dock, and then click the Splashtop Streamer disk image file. In the Finder window, double-click the Splashtop Streamer.pkg file to launch the Install Splashtop Streamer assistant, and then follow through the installation. You need to accept the license agreement, as usual, and then type your Administrator password for both the Splashtop Streamer installer and the Kextinstaller. When you finish the installation, Splashtop Streamer runs.
4. Next, you need to create the security code you will use to access Splashtop Streamer on your computer from your Kindle Fire or other devices. The code must be 8 to 20 characters long and must include at least one letter and at least one number.
   - **Windows**   The setup process displays the Create Security Code screen. Type your code in the Create Your Security Code box and the Confirm Your Security Code box, and then click the Next button.
   - **Mac**   The Splashtop Streamer window appears with the Security tab at the front, as shown here. Type your code in the Create Your Security Code box and the Confirm Your Security Code box. The Splashtop Streamer window displays the Status tab.

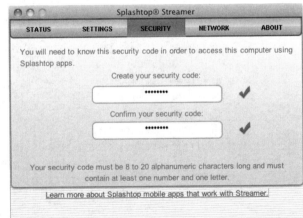

5. On Windows, the Done! screen now appears. Click the Finish button, and you'll see the Status tab of the Splashtop Streamer window, as shown here.

```
┌───────────────────────────────────────────────────┐
│                Splashtop® Streamer          [ _ ][ x ]│
├─────────┬──────────┬──────────┬──────────┬──────────┤
│ STATUS  │ SETTINGS │ SECURITY │ NETWORK  │  ABOUT   │
├─────────┴──────────┴──────────┴──────────┴──────────┤
│                                                     │
│   You can now use your mobile device to connect to this computer. │
│                                                     │
│                                                     │
│      98.143.144.103   Local Area Connection 2       │
│           10.0.0.48   Wireless Network Connection   │
│                                                     │
│                                                     │
├─────────────────────────────────────────────────────┤
│  Learn more about Splashtop mobile apps that work with Streamer. │
└─────────────────────────────────────────────────────┘
```

Install the Splashtop App on Your Kindle Fire

Now download and install the Splashtop app on your Kindle Fire. Follow these steps:

1. Tap the Apps button on the Home screen to display the Apps screen.
2. Tap the Store button to display the Store screen.
3. Tap the Search box, type **splashtop**, and then tap the Splashtop Remote Desktop search result.
4. On the search results screen, tap the Price button, and then tap the Buy App button.
5. When the installation finishes, tap the Open button to launch Splashtop.

Splashtop first shows brief instructions for installing and configuring the Splashtop Streamer app on your PC or Mac. Tap the Go button, the Continue button, and then the Finish button to cut to the chase. You'll then need to tap the Agree button on the End User License Agreement screen before you can start using Splashtop.

Connect Splashtop on Your Kindle Fire to Your Computer and Play Video

Next, Splashtop displays the Splashtop Remote Desktop screen, which lists the computers running Splashtop Streamer on the network. This illustration shows an example with a PC and a Mac.

Tap the button for your PC or Mac. Splashtop then displays the Splashtop Remote dialog box, shown here.

Type your security code, select the Remember Security Code check box if you want to store the code, and then tap the OK button.

Splashtop connects to your PC or Mac and automatically displays the Hints screen, shown here, which depicts the gestures you use on your Kindle Fire to take mouse actions on your computer. Clear the Show Hints Every Time check box when you've learned the gestures, and then tap the Continue button.

Your Kindle Fire's screen now displays the desktop on your PC or Mac (Figure 2-4 shows a Mac screen controlled by Splashtop), and you can navigate it by using gestures. For example:

- Tap and drag your finger to the left to scroll to the right.
- Place two fingers apart on the screen and pinch inward to zoom out. Place two fingers together on the screen and pinch apart to zoom in.
- Double-tap an item to click it.

Using your Kindle Fire, start a movie or video playing, and then resize the video window to fit the Kindle Fire's screen.

When you're ready to disconnect your Splashtop connection, tap the Menu button to display the Menu panel, tap the Back button, and then tap the OK button in the Disconnect dialog box (shown here).

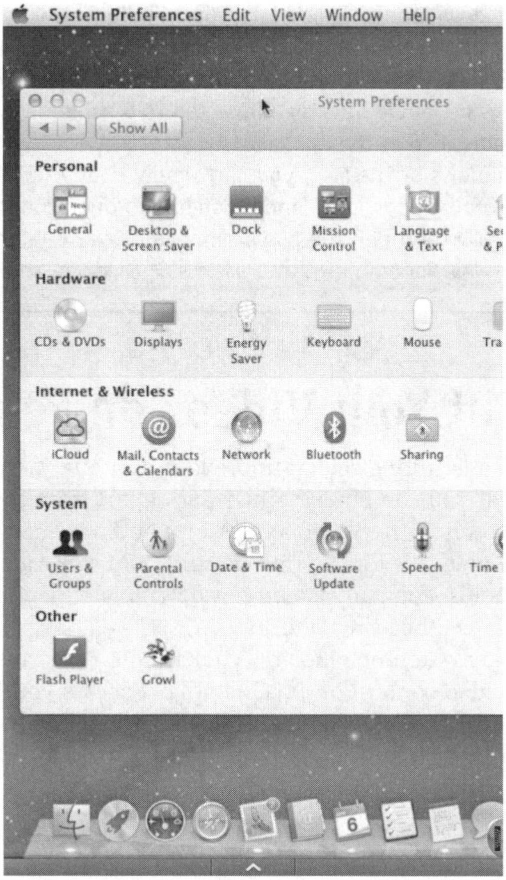

FIGURE 2-4   After you connect to your PC or Mac (shown here) with Splashtop, you can manipulate its user interface with your Kindle Fire—for example, to set a movie playing.

DOUBLE GEEKERY

Play a Movie from Your Kindle Fire on Your Computer

Unlike many other tablets and phones, the Kindle Fire doesn't have a way of outputting video to an external screen. So you can't connect your Kindle Fire to your TV and play your videos on the TV's screen, the way you can connect an iPad or a Samsung Galaxy Tab.

But if you've downloaded a movie to your Kindle Fire, you can play it on your computer easily enough. Connect your Kindle Fire to your computer so that the You Can Now Transfer Files From Your Computer To Kindle screen appears, and you can then either play the movie directly across the USB cable or copy the file to your computer and play it from there.

On either Windows or the Mac, you can simply double-click the movie file in a Windows Explorer window or a Finder window to open the movie in the default application. Or you can open the player you want to use—for example, the capable VLC (free from www.videolan.org) and use its Open command to open the movie.

Project 11: Put Your Videos on Your Kindle Fire

Amazon provides a tempting selection of video content, including TV series and full-length movies, that you can buy instantly and (apart from the cost) painlessly by using the Store screen in the Video app. And if you're a member of Amazon Prime (as every Kindle Fire owner is for the first month), you can watch videos you stream from a library of thousands. You can also buy or download video in Kindle Fire–compatible formats from various other sites online.

But if you enjoy watching video on your Kindle Fire, you'll almost certainly want to put your own video content on it. This project shows you how to do so. You may also want to rip files from your own DVDs so that you can watch them on your Kindle Fire. The next project explains how to do that.

DOUBLE GEEKERY

Learn What You Can and Can't Legally Do with Other People's Video Content

Before you start putting your videos and DVDs on your Kindle Fire, it's a good idea to know the bare essentials about copyright and decryption:

- If you created the video (for example, it's a home video or DVD), you hold the copyright to it, and you can do what you want with it—put it on your Kindle Fire, release it worldwide, or whatever. The only exceptions are if what you recorded is subject to someone else's copyright or if you're infringing on your subjects' rights (for example, to privacy).
- If someone has supplied you with a legally created video file that you can put on your Kindle Fire, you're fine doing so. For example, if you buy and download a movie from Amazon, you don't need to worry about legalities.
- If you own a copy of a commercial DVD, you need permission to rip (extract) it from the DVD and convert it to a format the Kindle Fire can play. Even decrypting the DVD in an unauthorized way (such as creating a file rather than simply playing the DVD) is technically illegal.

Create MP4 Files from Your Digital Video Camera

If you make your own movies with a digital video camera, you can easily put them on your Kindle Fire. To do so, you use an application such as Windows Live Movie Maker (Windows) or iMovie (Mac) to capture the video from your digital video camera and turn it into a movie file. Depending on which application you use to create the video, you may need to use another application to convert it to one of the video formats that the Kindle Fire plays.

Create MP4 Files Using
Windows Live Movie Maker or Windows Movie Maker

Unlike the last few versions of Windows, Windows 7 doesn't include Windows Movie Maker, the Windows program for editing videos. But you can download the nearest equivalent, Windows Live Movie Maker, from the Windows Live website (http://explore.live.com/windows-live-movie-maker).

 When you install Windows Live Movie Maker, the Windows Live Essentials installer encourages you to install all the Windows Live Essentials programs—Messenger, Photo Gallery, Mail, Writer, Family Safety, and several others. If you don't want the full set, click the Choose The Programs You Want To Install button on the What Do You Want To Install? screen, and then clear the check boxes for all the programs you don't want.

 DOUBLE GEEKERY

Understand Which Video Formats
Your Kindle Fire Can Play

Your Kindle Fire can play only two video formats:

- **MP4**   MPEG-4 Part 14, confusingly called MP4 in general use because the files use the .mp4 file extension, is part of the MPEG-4 video standard. MP4 can be used either for playing as a stream across the Internet or for playing from a file stored on your Kindle Fire.
- **VP8**   VP8 is an open format for video compression. VP8 was created by a company named On2 Technologies, but is now owned by Google.

In technical comparisons, MP4 and VP8 come out pretty much neck-and-neck—each has technical advantages and disadvantages that we won't get into here. But MP4 has what's called an *implementation advantage,* which means that there are many more tools for creating it, and so many more people are using it.

So normally, you'll want to create files in the MP4 format rather than the VP8 format. This chapter shows you how to create MP4 files.

Windows Live Movie Maker can't export video files in MP4 format, so what you need to do is export the video file in the WMV format, and then convert it using another application, such as Full Video Converter Free (discussed later in this chapter).

Similarly, the versions of Windows Movie Maker included with Windows Vista and Windows XP can't export MP4 video files, so you need to export the video file in a standard format (such as AVI) that you can then convert using another application.

Create a WMV File from Windows Live Movie Maker   To create a WMV file from Windows Live Movie Maker, open the project and follow these steps:

1. Click the unnamed tab at the left end of the Ribbon to display its menu, and then click the Save Movie item to display the Save Movie panel.
2. In the Common Settings section, click For Computer. The Save Movie dialog box opens.
3. Type the name for the movie, choose the folder in which to store it, and then click the Save button.

Now that you've created a WMV file, use a converter program such as Full Video Converter Free (discussed later in this chapter) to convert it to a format that your Kindle Fire can play.

Create an AVI File from Windows Movie Maker on Windows Vista   To save a movie as an AVI file from Windows Movie Maker on Windows Vista, follow these steps:

1. With your movie open in Windows Movie Maker, choose File | Publish Movie (or press CTRL-P) to launch the Publish Movie Wizard. The Wizard displays the Where Do You Want To Publish Your Movie? screen.
2. Select the This Computer item in the list box, and then click the Next button. The Wizard displays the Name The Movie You Are Publishing screen.
3. Type the name for the movie, choose the folder in which to store it, and then click the Next button. The Wizard displays the Choose The Settings For Your Movie screen (see Figure 2-5).
4. Select the More Settings option button, and then select the DV-AVI item in the drop-down list.

 The DV-AVI item appears as DV-AVI (NTSC) or DV-AVI (PAL), depending on whether you've chosen the NTSC option button or the PAL option button on the Advanced tab of the Options dialog box. NTSC is the video format used in most of North America; PAL is the format used in Europe, much of Asia, and Australia.

5. Click the Publish button to export the movie in this format. When Windows Movie Maker finishes exporting the file, it displays the Your Movie Has Been Published screen.

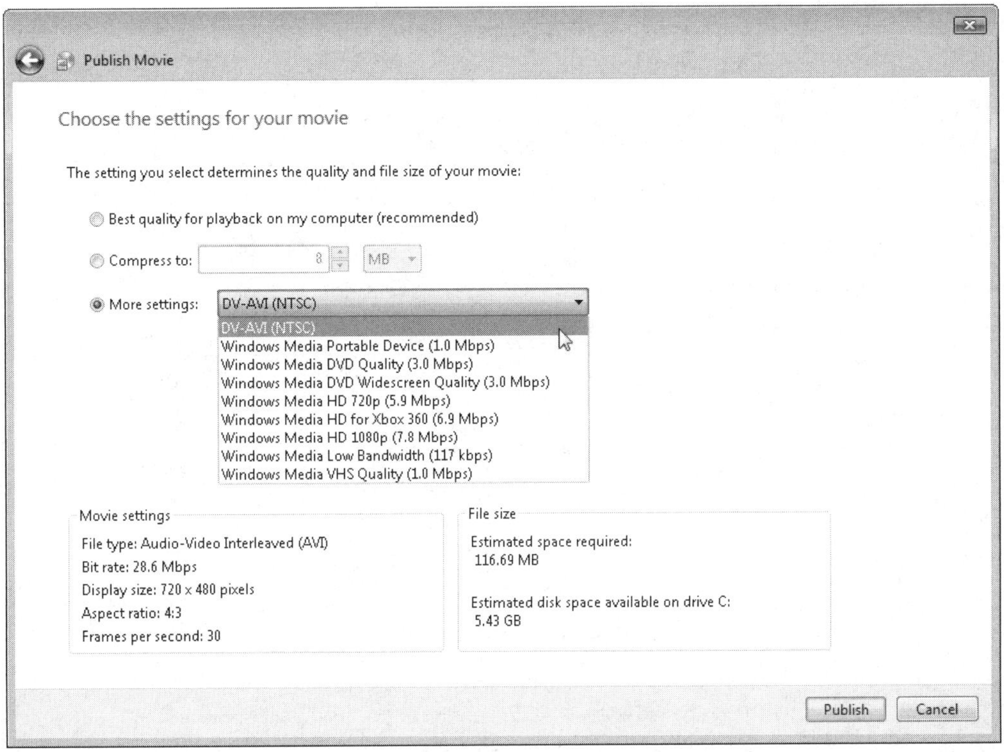

FIGURE 2-5   On the Choose The Settings For Your Movie screen, select the More Settings option button, and then pick the DV-AVI item in the drop-down list.

6. Clear the Play Movie When I Click Finish check box if you don't want to watch the movie immediately in Windows Media Player. Often, it's a good idea to check that the movie has come out okay.
7. Click the Finish button.

Now that you've created an AVI file, use a converter program such HandBrake (discussed later in this chapter) to convert it to a format that works on your Kindle Fire.

Create an AVI File from Windows Movie Maker on Windows XP   To save a movie as an AVI file from Windows Movie Maker on Windows XP, follow these steps:

1. Choose File | Save Movie File to launch the Save Movie Wizard. The Wizard displays its Movie Location screen.
2. Select the My Computer item, and then click the Next button. The Wizard displays the Saved Movie File screen.
3. Enter the name and choose the folder for the movie, and then click the Next button. The Wizard displays the Movie Setting screen (shown in Figure 2-6 with options selected).

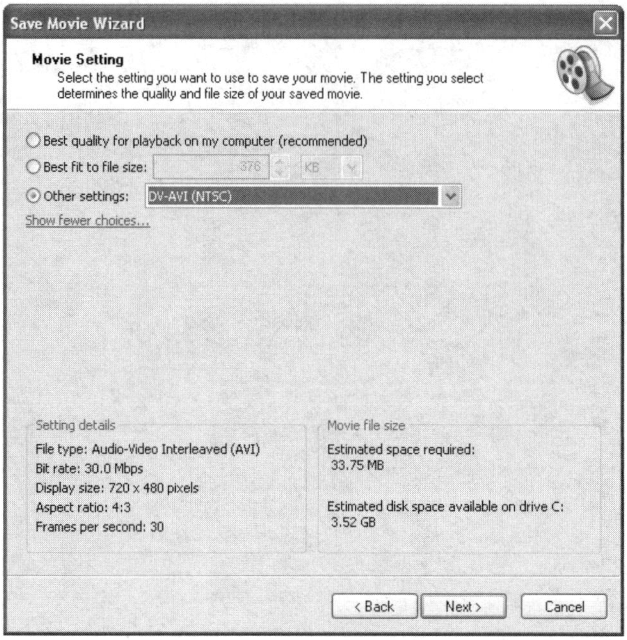

Save Movie Wizard

Movie Setting
Select the setting you want to use to save your movie. The setting you select
determines the quality and file size of your saved movie.

○ Best quality for playback on my computer (recommended)
○ Best fit to file size: 376   KB
◉ Other settings: DV-AVI (NTSC)
Show fewer choices...

Setting details

File type: Audio-Video Interleaved (AVI)
Bit rate: 30.0 Mbps
Display size: 720 x 480 pixels
Aspect ratio: 4:3
Frames per second: 30

Movie file size

Estimated space required:
33.75 MB

Estimated disk space available on drive C:
3.52 GB

[< Back] [Next >] [Cancel]

FIGURE 2-6   On the Movie Setting screen, select the
Other Settings option button and pick the DV-AVI item from
the drop-down list.

4. Click the Show More Choices link to display the Best Fit To File Size option
   button and the Other Settings option button.
5. Select the Other Settings option button, and then select the DV-AVI item in the
   drop-down list.

The DV-AVI item appears as DV-AVI (NTSC) or DV-AVI (PAL), depending on whether
you've chosen the NTSC option button or the PAL option button on the Advanced
tab of the Options dialog box. NTSC is the video format used in most of North
America; PAL's stronghold is Europe.

6. Click the Next button to save the movie in this format. The Wizard displays the
   Completing The Save Movie Wizard screen.
7. Clear the Play Movie When I Click Finish check box if you don't want to test the
   movie immediately in Windows Media Player. Usually, it's a good idea to make
   sure the movie has come out right.
8. Click the Finish button.

Now that you've created an AVI file, use a converter program such as Full Video Converter Free (discussed later in this chapter) to convert it to a format that works on your Kindle Fire.

Create MP4 Files Using iMovie

To use iMovie to create video files that will play on your Kindle Fire, follow these steps:

1. With the movie open in iMovie, choose Share | Export Using QuickTime. iMovie displays the Save Exported File As dialog box (shown here).

> | | |
> |---|---|
> | Save exported file as... | |
> | Save As: | Kiting Seminar.mp4 |
> | Where: | 📁 Movies |
> | Export: | Movie to MPEG-4 | Options... |
> | Use: | Default Settings |
> | | Cancel   Save |

2. In the Save As box, type the name under which you want to export the movie.
3. In the Where pop-up menu, choose the folder in which to store the exported movie. If necessary, click the down arrow to the right of the Save As box to expand the dialog box so that you can navigate through the file system to another folder.

 If your Kindle Fire is connected to your Mac, you can export the video file directly to the Kindle Fire's Video folder.

4. In the Export pop-up menu, choose Movie To MPEG-4.
5. Click the Options button to display the MPEG-4 Export Settings dialog box (see Figure 2-7).
6. In the File Format pop-up menu at the top, choose MP4.
7. Make sure the Video tab is selected. If not, click it.
8. In the Video Format pop-up menu, choose MPEG-4 Improved.
9. In the Image Size pop-up menu, choose 1280 × 720 HD.
10. In the Frame Rate pop-up menu, choose the frame rate you want. Normally this will be 24 for NTSC video and 30 for PAL video.
11. Click the OK button to close the MPEG-4 Export Settings dialog box.
12. Click the Save button to save the file. You'll see the Exporting Project dialog box while iMovie exports the file.

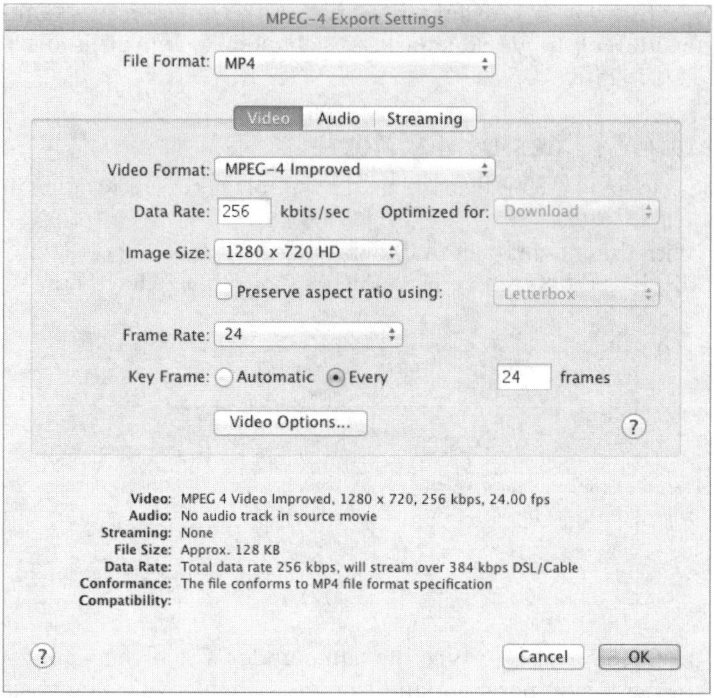

FIGURE 2-7   In the MPEG-4 Export Settings dialog box, choose
MPEG-4 Improved in the Video Format pop-up menu and 1280 × 720
HD in the Image Size pop-up menu.

DOUBLE GEEKERY

Play Video Files in Other Formats
on Your Kindle Fire

As you learned earlier in this chapter, your Kindle Fire's built-in Videos app and
Gallery app can play only two video formats: MP4 and VP8. So if you have video
files in other video formats, such as AVI or MOV, you'll need to convert them to MP4
before you can play them in the Videos app or the Gallery app.

As you'll see in the rest of this project, you can convert video files by using your
computer. But if you prefer to simply work around the problem, try installing the
MoboPlayer app instead. It can play many video file types, including AVI and MOV.

At this writing, MoboPlayer isn't available from the Appstore, so you'll need to
side-load it. Download the MoboPlayer package from one of the sources explained in
Project 33, and then side-load the app as explained in Project 34 (both in Chapter 4).

After installing MoboPlayer, tap the Open button to open the app. You see
the opening screen (shown next). You can tap the Scan SDCard button to scan
your Kindle Fire's sdcard folder (whose contents include the Videos folder), tap

the Intelligent Scanning button to have MoboPlayer scan the folders it expects to contain video files, or simply tap the Directory Browsing button to browse the directories on your own.

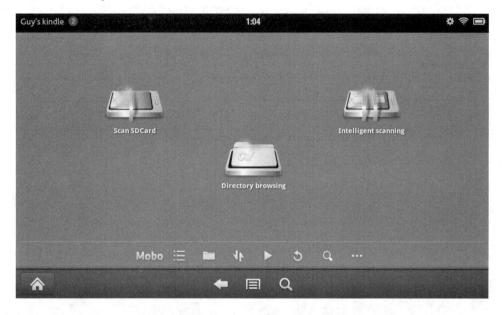

Whichever method you use, tap the movie you want to view. MoboPlayer opens the movie and starts it playing. You can control playback by using the controls that pop up at the bottom of the screen when you tap the screen.

Create MP4 Video Files from Your Existing Video Files

If you have video files in other formats than MP4 (for example, files in the AVI format or QuickTime movies), you will need to convert them to MP4 format before you can play them on your Kindle Fire.

The easiest way to convert the files is by using Handbrake, a free application you can download from the Internet. Handbrake can open files in many video formats and convert them to other formats, including MP4 format. Even better, you can specify the output size and quality, so you can choose the best balance between video quality and compact file size.

Download, Install, and Launch HandBrake

First, install HandBrake on your PC or Mac. Follow these steps:

1. Open your web browser and go to the Download page on the HandBrake website (http://handbrake.fr/downloads.php).
2. Click the appropriate download link under the Mac OS heading or the Windows heading.

 DOUBLE GEEKERY

Learn About Other Tools for Converting Video Files to MP4 Files

HandBrake is a great tool for converting video files from various formats to MP4 files. But if you have files that HandBrake can't convert, you may need to use other tools. Here are three possibilities:

- **Full Video Converter Free**   Full Video Converter Free is a freeware Windows program you can download from Top 10 Download (www.top10download.com) and other sites. When you install the program, make sure you decline any extra options such as adding a toolbar, changing your default search engine, or changing your home page.
- **Kindle Fire Video Converter**   The Kindle Fire Video Converter website (http://kindlefirevideoconverter.com) offers both Windows and Mac versions of its Video Converter app. Each costs $49.99, but there's an evaluation version you can try for 30 days. The evaluation version puts a watermark on the videos you convert with it.
- **Zamzar**   Zamzar (www.zamzar.com) is an online file conversion tool. For low volumes of files, the conversion is free (though it may take a while), but you must provide a valid e-mail address. For higher volumes of files or higher priority, you can sign up for a paid account.

You can find various other free programs online for converting video files. If you're looking for such programs, check carefully that what you're about to download is actually free rather than a crippled version that requires you to pay before you can convert files.

DOUBLE GEEKERY

Find Out Whether Your Version of Windows Is 32-Bit or 64-Bit

HandBrake comes in different versions for Windows—a 64-bit version and a 32-bit version. You'll likely need the 32-bit version, because at this writing most Windows PCs run 32-bit versions of Windows; if you're lucky enough to have a PC running 64-bit Windows, you'll probably know about it.

Sixty-four-bit versions of Windows XP are rare, especially nowadays, so you'll only really need to check on Windows 7 or Windows Vista. To check, press WINDOWS KEY–BREAK to display the System window (on Windows 7 or Windows Vista) or the System Properties dialog box (on Windows XP). Then look at the System Type readout in the System section to see if it says 32-Bit Operating System or 64-Bit Operating System.

3. When the download finishes, install the application as usual:
   - **Windows** Click the Run button to launch the HandBrake Setup Wizard, accept the license agreement and the default settings, and then wait while the Wizard installs HandBrake.
   - **Mac** If OS X doesn't automatically open a Finder window showing the contents of the HandBrake disk image, click the Downloads icon on the Dock, and then click the HandBrake disk image file. In the Finder window, drag the HandBrake icon to the Applications folder. When you launch HandBrake, you'll need to accept the license agreement.

Now that you've installed HandBrake, launch it:

- **Windows** Choose Start | All Programs | HandBrake | HandBrake.
- **Mac** Click the Launchpad icon on the Dock, and then click the HandBrake icon. If your version of OS X doesn't have Launchpad, click the desktop, choose Go | Applications, and then double-click the Finder icon in the Applications folder.

Figure 2-8 shows the HandBrake window with a file loaded and the Presets pane open on the right.

 This section shows HandBrake on the Mac. On Windows, HandBrake is almost exactly the same, but I'll point out the differences you need to know about.

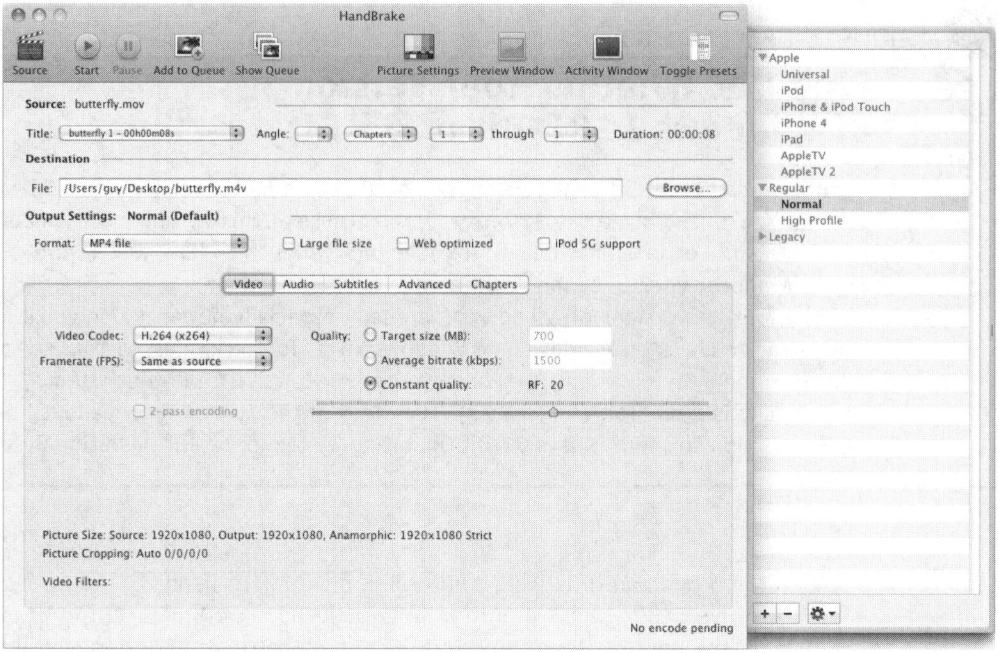

FIGURE 2-8   The Presets pane on the right side of the HandBrake window lets you instantly choose video settings for your Kindle Fire.

Convert Video Files to MP4 Files Using HandBrake

With HandBrake open, you're ready to start converting files. Follow these steps:

1. Display the Open dialog box:
   - **Windows**   Click the Source drop-down button on the toolbar, and then click Video File. You can also press CTRL-O.
   - **Mac**   Click the Source button on the toolbar.
2. Click the file you want, and then click the Open button. HandBrake reads the file and then displays its details.
3. In the File box in the Destination area, specify the folder and filename to use for the converted file. You can either type in the folder path or click the Browse button and select it in the dialog box that opens.
4. If the Presets pane isn't displayed, click the Toggle Presets button on the toolbar to display it. Then click the preset you want to use.

On Windows, you can choose the output picture size and cropping by using the controls on the Picture tab in the lower part of the window.

⚙ DOUBLE GEEKERY

Create a Preset for the Kindle Fire

To get the best video files for your Kindle Fire, you'll need to use custom settings. You can change the settings with only a little effort, but what you should do is create a preset so that you can apply the settings quickly and consistently. At this writing, you can create this kind of preset, with a custom picture size, only on the Mac.

At this writing, HandBrake doesn't have a preset for the Kindle Fire, so you need to create one. But that may have changed by the time you read this, so check whether your version of HandBrake has such a preset. If the Presets pane isn't displayed on the right side of the window, click the Toggle Presets button on the toolbar to display it. Then expand the Devices category and look for a Kindle Fire entry. If there is one, test it and see if it meets your needs.

To create a preset, follow these steps:

1. Click the Add button (on the Mac, the + button) at the bottom of the Presets pane. HandBrake displays the Add New Preset dialog box (shown here).
2. In the Preset Name text box, type a descriptive name, such as **Kindle Fire.**
3. Open the Use Picture Size drop-down list and choose Custom; then enter **1024** in the first box and **600** in the second box.
4. Type a description of the preset in the Description box if you will find it helpful.
5. Click the Add button. Your new preset appears at the bottom of the list in the Presets pane.

| Preset Name: |
| Kindle Fire |

Picture Settings:
Use Picture Size: Custom
1024 X 600

Use Picture Filters: ☐

Description:
Full-screen video for the Kindle Fire

Cancel   Add

5. Open the Container drop-down list on Windows or the Format pop-up menu on the Mac, and then choose MP4 File.
6. Select the Large File Size check box if you're okay with a relatively large file size. This reduces the amount of encoding, so it's worth experimenting with—but the Kindle Fire's small amount of storage will probably lead you to prefer smaller files.
7. Make sure the Video tab is selected in the lower part of the window. If not, click the Video tab.
8. In the Video Codec drop-down list, choose H.264 (x264).

9. In the Framerate (FPS) drop-down list, choose Same As Source.
10. Click the Start button on the toolbar. HandBrake starts converting the file. When it finishes, HandBrake displays the dialog box shown here.

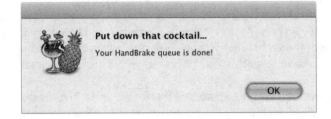

Project 12: Put Your DVDs on Your Kindle Fire

If you have DVDs, you'll probably want to put them on your Kindle Fire so that you can watch them anytime and anywhere. This project gives you an overview of how to create suitable files, first on Windows, and then on the Mac.

 Because ripping commercial DVDs without specific permission is a violation of copyright law, there are no DVD ripping programs from major companies. You can find commercial programs, shareware programs, and freeware programs on the Internet—but keep your wits firmly about you, as some programs are a threat to your computer through being poorly programmed, while others include unwanted components such as adware or spyware. Always read reviews of any DVD ripper you're considering before you download and install it—and certainly before you pay for it. As usual on the Internet, if something seems too good to be true, it most likely *is* too good to be true.

Before you start ripping, make sure that your discs don't contain computer-friendly versions of their contents. At this writing, some Blu-Ray discs include such versions, which are licensed for you to load on your computer and your lifestyle devices (such as your Kindle Fire).

 DOUBLE GEEKERY

Understand What DVD Titles and Chapters Are

Each DVD is split up into titles and chapters:

- **Title** A *title* is one of the recorded tracks on the DVD.
- **Chapter** The *chapters* are the bookmarks within the titles—for example, if you press the Next button on your remote, your DVD player skips to the start of the next chapter.

Rip DVDs with DVDFab HD Decrypter on Your PC

In this section, we'll install DVDFab HD Decrypter on your PC and use it to rip DVDs. DVDFab HD Decrypter is shareware that you can try for 30 days, which gives you plenty of time to decide whether it does what you want. After that, you're supposed to pay to register it.

Install DVDFab HD Decrypter on Your PC

To download and install DVDFab HD Decrypter on your PC, open your web browser and go to the DVDFab HD Decrypter page on the DVDFab.com website, www.dvdfab

.com/hd-decrypter.htm. Click the Download button, click the Run button in the File Download – Security Warning dialog box, and then click the Run button in the Internet Explorer – Security Warning dialog box (shown here).

The installer then runs. Follow through the installation as usual. There are three points worth noting:

- In the Select Setup Language dialog box, choose your language, and then click the OK button.
- On the Select Additional Tasks screen (shown here), clear the Create A Desktop Icon check box if you don't want to create a desktop icon for DVDFab HD Decrypter. Similarly, clear the Create A Quick Launch Icon check box if you don't want to create an icon on the Quick Launch toolbar.

- On the Completing The DVDFab 8 Qt Setup Wizard, select the Yes, Restart The Computer Now option button if you're okay with restarting immediately. Otherwise, select the No, I Will Restart The Computer Later option button and restart at your convenience.

Rip a DVD with DVDFab HD Decrypter

After restarting your PC, log in as usual. You'll then be ready to rip a DVD with DVDFab HD Decrypter. Follow these steps:

1. Launch DVDFab HD Decrypter from the Start menu (choose Start | All Programs | DVDFab 8 Qt | DVDFab 8 Qt), the desktop shortcut, or the Quick Launch toolbar.
2. Insert a DVD in your PC's DVD drive, and wait while DVDFab HD Decrypter identifies its contents.
3. Click the DVD Ripper button in the left pane of the DVDFab window to display the DVD ripping options (see Figure 2-9).
4. Click the More button to display the pop-up menu of devices and formats, and then click the kindlefire item. The kindlefire button appears in the DVD Ripper box, as shown here.
5. In the Target text box, enter the folder in which to store the ripped file. You can type in the path, click the button with the folder icon, and use the Please Choose A Folder dialog box to select the folder; or you can click the drop-down button and use a folder you've used before.

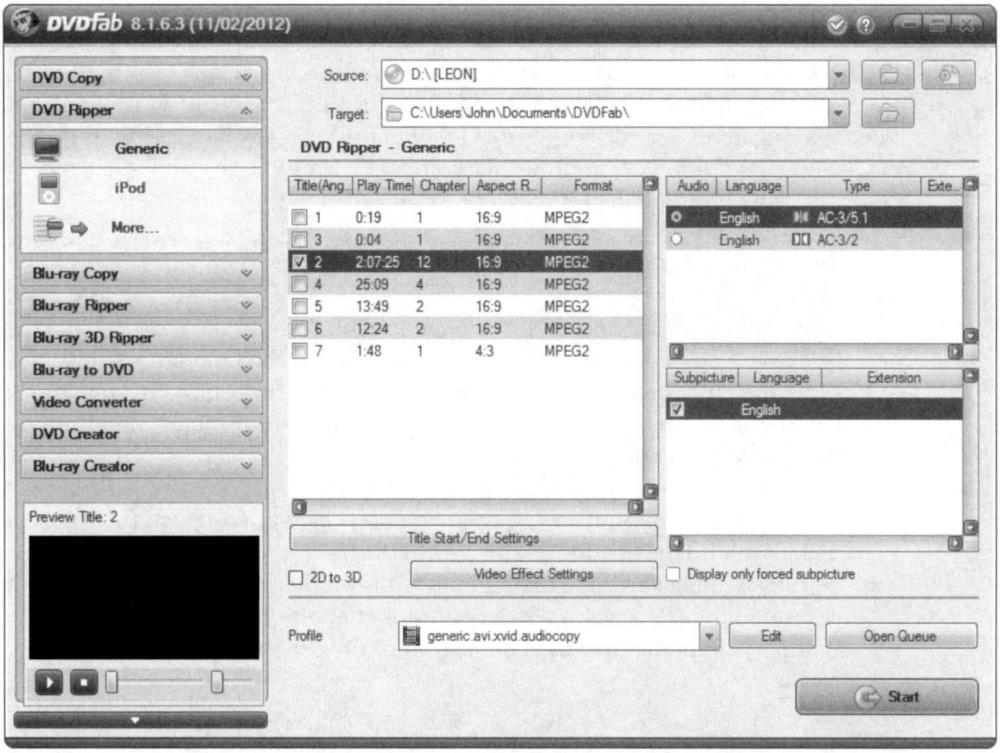

FIGURE 2-9   Click the DVD Ripper button in the left pane to display the DVD ripping options; then click the More button to display the pop-up menu of devices and formats.

6. In the list of titles in the center of the window, select the check box for the DVD title you want to rip. Normally, you'll want the DVD's main title, which you can pick out easily by its length—it's the length of the movie.

7. If you want to rip only some chapters from the title you've chosen, click the Title Start/End Settings button to display the Title Start/End Settings dialog box (shown here). Choose the chapters by using the Start drop-down list and End drop-down list, and then click the OK button.

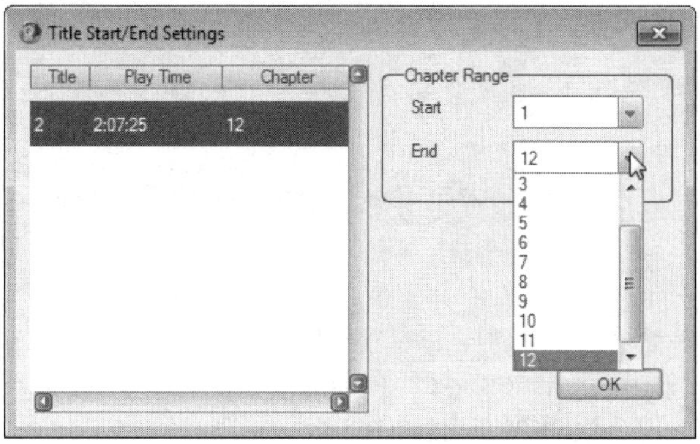

8. Click the Start button to start ripping. When DVDFab HD Decrypter finishes ripping the file, it opens a Windows Explorer window to the folder containing the file.

Rip DVDs with HandBrake on Your Mac

The best tool for ripping DVDs on the Mac is HandBrake, which you met in the previous project. To rip DVDs with HandBrake, you must install VLC, a DVD- and video-playing application (free; www.videolan.org). This is because HandBrake uses VLC's DVD-decryption capabilities; without VLC, HandBrake cannot decrypt DVDs.

Once you've installed VLC, you can rip DVDs like this:

1. Run HandBrake as usual. For example, click the Launchpad icon on the Dock, and then click the HandBrake icon, or open a Finder window to your Applications folder and then double-click the HandBrake icon.
2. Click the Source button on the toolbar to display the Open dialog box.
3. Click the DVD in the Source list on the left, and then click the Open button. HandBrake scans the DVD, which may take several minutes, and then displays its details (see Figure 2-10).
4. In the Title pop-up menu, choose which title to rip. To rip a movie from a DVD, you'll want the main title. Usually, you can easily distinguish the main title by its length—it'll be as long as the movie (for example, two hours).
5. Optionally choose which chapters to rip from the movie by using the Chapters pop-up menus. For example, if you want chapters 1 through 10, choose 1 in the first pop-up menu, and then choose 10 in the second pop-up menu.

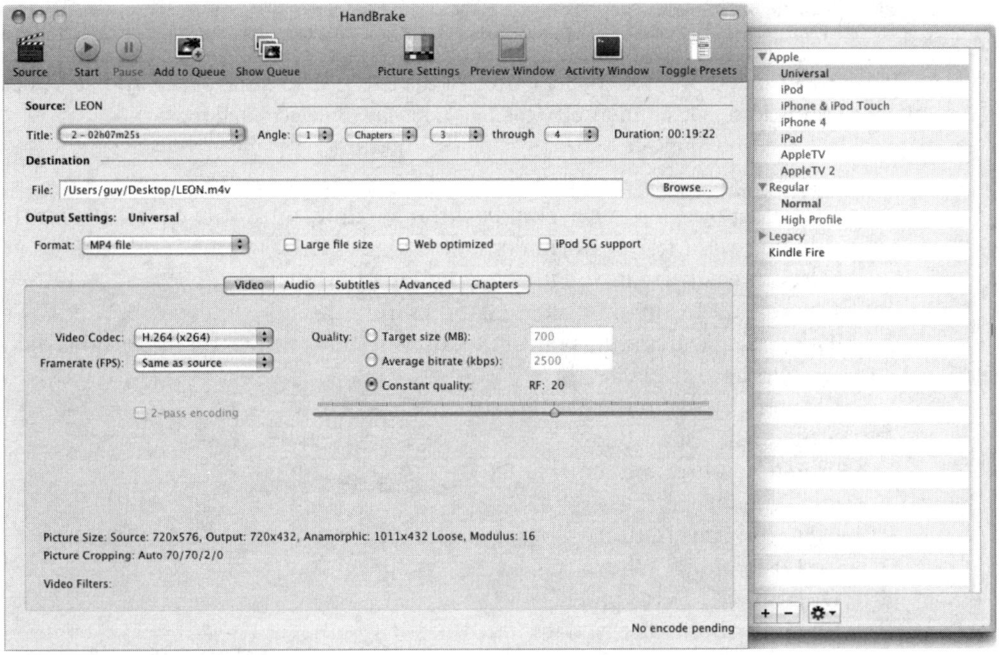

FIGURE 2-10 HandBrake scans the DVD and displays its details. You can then choose which title and chapters to rip.

If the DVD offers multiple angles, open the Angle pop-up menu and choose the angle you want.

6. In the Destination box, enter the folder and filename to use for the video file. You can type the path and filename if you want, but it's usually easier to click the Browse button, use the Save dialog box that opens to specify the folder and filename, and then click the Save button.

7. If the Presets pane isn't displayed on the right side of the HandBrake window, click the Toggle Presets button to display it.

8. Click the preset you want to use. For example, if you created a Kindle Fire preset as recommended in the previous project, click that preset. The settings in the lower part of the window change to the settings for the preset.

9. In the Output Settings area, select MP4 File in the Format pop-up menu. Normally, you'll want to make sure the Large File Size check box, the Web Optimized check box, and the iPod 5G Support check box are all cleared.

10. In the lower part of the HandBrake window, choose suitable settings on the Video tab. These are the settings that work best for video files you'll play on the Kindle Fire:
    - **Video Codec**   Choose H.264 (x264).
    - **Framerate (FPS)**   Choose Same As Source.
    - **Quality**   Select the Constant Quality option button, and then drag the slider to the position you find gives you the quality you want. You'll need to experiment with this setting by creating video files at different qualities and seeing which look good enough on your Kindle Fire.

11. Click the Start button on the toolbar to start ripping.

 DOUBLE GEEKERY

Prevent the Mac OS X DVD Player from Running Automatically When You Insert a DVD

When you insert a movie DVD, Mac OS X automatically launches DVD Player, switches it to full screen, and starts the movie playing. This behavior is great for when you want to watch a movie, but not so great when you want to rip it.

To prevent DVD Player from running automatically when you insert a DVD, follow these steps:

1. Choose Apple | System Preferences to open System Preferences.
2. In the Hardware section, click the CDs & DVDs item.
3. In the When You Insert A Video DVD drop-down list, you can choose Ignore if you want to be able to choose freely which application to use each time. If you always want to use the same application, choose Open Other Application, use the resulting Open dialog box to select the application, and then click the Choose button.
4. Choose System Preferences | Quit System Preferences or press ⌘-Q to close System Preferences.

Project 13: Use Your Kindle Fire for In-Car Entertainment

If you drive your car, the road and its denizens probably give you all the visual entertainment you need—and you certainly shouldn't be using your Kindle Fire at the wheel when you're busy drinking your coffee and applying your nail polish.

But your passengers may well need extra entertainment—and the Kindle Fire is a great way to give them personalized entertainment programs and keep them off your back until you reach your destination.

In a pinch, a passenger can simply hold the Kindle Fire—no extra equipment required. But for that airline in-seat entertainment touch, you'll probably want to get a car headrest mount holder like the one shown in Figure 2-11.

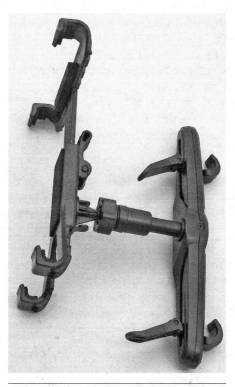

FIGURE 2-11 An inexpensive car headrest mount holder like this is all you need to mount a Kindle Fire on the rear of a headrest to keep a rear-seat passenger entertained.

There are two main things to check when choosing a car headrest mount holder. First, make sure that it will hold the Kindle Fire snugly. Some holders designed for heftier tablets, such as the iPad and the Samsung Galaxy Tab, only just go small enough for the Kindle Fire. Second, make sure it's robust enough to hold the Kindle Fire still even as you thunder along dilapidated highways. Some mounts are pretty feeble. You can improve them with duct tape, but ideally you'll buy a holder that's strong enough in the first place.

Once you've gotten your car headrest mount holder, mounting the Kindle Fire takes only a minute or two. Figure 2-12 shows a Kindle Fire set up in portrait mode for reading on the road. If your passengers will play movies instead, mount the Kindle Fire in landscape mode.

Now your passenger can plug in her headphones, set her preferred movie playing, and you'll be ready to roll.

FIGURE 2-12 Using the holder, you can quickly mount the Kindle Fire to the back of the headrest. This headrest has a rotating joint that enables you to switch quickly between portrait and landscape orientations and to adjust the Kindle Fire's angle in either orientation.

3 Books Geekery

As you've seen in the previous two chapters, the Kindle Fire is great for playing back music and video. But it's arguably even better for reading books: not only is the screen the right size to display the average page, but you can load hundreds or even thousands of books in the Kindle Fire's storage space and still have room for your songs and videos.

In this chapter, I'll show you how to make the most of books on the Kindle Fire. I bet you already know how to open, navigate, and read books using the Books app, so we'll concentrate on the other part of the problem—getting the books you want without paying the earth for them, converting books from other formats into the Kindle Fire's preferred format, and making your Kindle Fire realize they're books rather than documents.

We'll start by going quickly through the different e-book formats in use, identifying those your Kindle Fire can display and pointing out those it can't handle. We'll then look at sources of free e-books before talking about ways of adding your existing e-books to your Kindle Fire. We'll then grab and install the free e-book manager called calibre and learn its essentials before using it and other tools to convert books to Kindle Fire–friendly formats.

Toward the end of the chapter, I'll show you how to read your e-books in the bath safely (or watch videos in the mud bath if you prefer). Finally, I'll show you how to read your Kindle books on your computer—or on someone else's computer.

Project 14: Understand Which E-book Formats Your Kindle Fire Can Use

E-books come in many different file formats, each of which has its advantages and disadvantages. If you simply buy all your books from Amazon, you'll avoid most of the problems caused by these differing file formats—but chances are that you'll want to get at least some of your books from other sources. That means it's a good idea to know which file formats are good for your Kindle Fire, which you can easily convert to make them Kindle Fire–friendly, and which to avoid.

Table 3-1 lists the e-book file formats your Kindle Fire can display.

Table 3-2 lists the four most widely used e-book file formats your Kindle Fire can't display.

TABLE 3-1 E-book File Formats Your Kindle Fire Can Display

Format	File Extension	Explanation	Comments
Mobipocket	.mobi .prc	An e-book format based on the Open eBook standard. Mobipocket was a French company launched in 2000. Amazon bought Mobipocket in 2005.	Your Kindle Fire can read Mobipocket files natively. Files with either the .mobi file extension or the .prc file extension should display on your Kindle Fire with no problems.
Amazon Kindle	.azw	Amazon's proprietary Kindle Format. This format is based on the Mobipocket standard, but uses different formatting and adds digital rights management (DRM).	This is the original Kindle format. E-books in this format can be protected with DRM to prevent you from sharing them with others.
Kindle Format 8	.kf8	Amazon's next-generation file format for Kindles, including the Kindle Fire. Improvements to the file format include support for HTML 5 and CSS 3, the latest standards for formatting webpages.	This is the best format for the Kindle Fire, because it gives the greatest variety of content and control over it. Older Kindles can't read e-books in this format.
Text	.txt	Plain text files—text only, without formatting, pictures, or other objects.	You can put these on your Kindle Fire easily enough, but the text looks pretty boring without formatting or layout.
Portable Document Format	.pdf	Adobe's industry-standard format for sharing documents in a laid-out format. Some PDF documents are marked as being reflowable, so that devices with different-size screens can display them effectively. Other PDF documents are not reflowable.	Reflowable PDFs work well on the Kindle Fire's screen. Nonreflowable PDFs are difficult to read, because you often need to scroll around and zoom in and out to see the whole of a page.
Microsoft Word	.docx .doc	The proprietary format for Microsoft Word documents. Word 2007 and later versions use the .docx format natively but can also create .doc files. Word 2004 and earlier versions use the .doc format.	Your Kindle Fire reformats and reflows the contents of a Word document to fit the screen so that it's easy to read. This works well for text-based documents, but complex elements such as tables and graphics may be hard to view.

TABLE 3-2 E-book File Formats Your Kindle Fire Can't Display

Format	File Extension	Explanation	Comments
EPUB	.epub	An open e-book standard that was created by the International Digital Publishing Forum (IDPF).	Many publishers produce ePub files, and most e-readers can display them—but the Kindle Fire can't at this writing. These are the files you're most likely to need to convert.
eReader	.pdb	The file format for Palm Digital Media e-books.	eReader files were introduced when Palm devices were popular. In 2009, Barnes & Noble announced that it would use eReader files for the e-books it sells.
Microsoft Reader	.lit	The proprietary file format for Microsoft Reader. These files can be protected with DRM.	Microsoft plans to discontinue Microsoft Reader in August 2012.
Newton eBook	.pkg	The file format used for Apple's Newton tablet, a 1990s PDA. Each file is a package that can contain multiple books.	Newton eBook files are relatively rare except for some specialized genres such as sci-fi.

Looking at Tables 3-1 and 3-2, you'll quickly see that EPUB is the 800-pound gorilla that Amazon wishes would leave the e-book jungle. EPUB is widely used and works well for many publishers, but Amazon has chosen not to support it on the Kindle.

There's no technical reason why the Kindle Fire can't read EPUB files—almost every other e-reader can, including e-readers that are much less powerful and capable than the Kindle Fire—but there are plenty of marketing reasons.

Because of this situation, we'll spend a fair part of this chapter looking at how to get EPUB files converted so that they'll work well on your Kindle Fire.

 DOUBLE GEEKERY

Deal with Books Disappearing from the Cloud Screen or the Device Screen

If some of your books disappear from either the Cloud screen or the Device screen in the Books app, follow these steps to make them appear again:

1. Tap the Settings button on the status bar to display the settings bar.
2. Make sure Wi-Fi is on. If not, tap it, tap the Wireless Networking switch, and connect to one of your wireless networks.
3. Tap the More button to display the Settings screen.

4. Tap the Applications button to display the Applications screen.
5. In the Filter By pop-up menu, select All Applications.
6. Tap the Amazon Kindle button to display the Amazon Kindle settings screen.
7. In the Storage area, tap the Clear Data button.
8. Tap the Home button to return from the Settings app to the main screen.

After you clear the data like this, your Kindle Fire refreshes the data for the books on both the Cloud screen and the Device screen. Your missing books then appear along with all your other books.

Project 15: Get Free E-books Instead of Buying E-books

Before buying any e-books, be sure to make the most of the free e-books you can get on the Internet. You can load up on e-book versions of classics in just a few minutes at no cost, but you can also find contemporary fiction for free. And if you subscribe to the Amazon Prime service, you can borrow an e-book each month from the Kindle Owners' Lending Library.

This project points you to six great sources of free books.

Free Books on Amazon.com

As well as the millions of books you can buy, Amazon.com has a good range of free books—but you have to find them.

Visit the Top 100 Free List on Your Kindle Fire

When you're using your Kindle Fire, the best way to get started with free books is to visit the Top 100 Free list. Follow these steps:

1. From the Home screen, tap the Books button to display the Books app.
2. Tap the Store button to display the Store screen.
3. Scroll down to the bottom of the screen, where you'll find the Top 100 Free section.
4. Tap the See All button. The Top 100 Free list appears.

 Many publishers give away some free copies of a new book to seed the market, but then restore the book's price when it becomes more popular. So if you hear a new book is free, grab it before the price has a chance to change.

Visit the Kindle Popular Classics Section on Your Computer

Amazon's Kindle Popular Classics section has more than 15,000 free classic books at this writing, including works from writers such as Jane Austen, Charles Dickens, and Thomas Paine.

To find the Kindle Popular Classics section, open your web browser and go to www.amazon.com/s/?node=2245146011. You can then browse to find books that interest you or search by author or title.

 You can also browse the Kindle Popular Classics section by entering the search term **kindle popular classics** in the Books app on your Kindle Fire. But when you do this, you get only a much shorter list of books—fewer than 1000 at this writing.

Kindle Owners' Lending Library

The Kindle Owners' Lending Library is part of Amazon.com, a feature for Amazon Prime members. You can borrow one Kindle book for free each month. At this writing, the Kindle Owners' Lending Library offers more than 50,000 books, so there's a good chance it'll have some books you want to read.

 Amazon Prime costs $79 per year at this writing and gives you free two-day shipping on orders from Amazon, unlimited streaming of those thousands of movies and TV shows available for streaming, and one free book rental a month. When you sign up as a new Kindle Fire user, you get a one-month trial of Amazon Prime. Make the most of the trial to help decide whether you want to pay for more.

As you'd guess, you use your Kindle Fire to browse the Kindle Owners' Lending Library and borrow books. To access the Kindle Owners' Lending Library, follow these steps:

1. Tap the Home button to display the Home screen.
2. Tap the Books button on the bar at the top to display the Books app.
3. Tap the Store button in the upper-right corner to display the Store.
4. In the lower-right pane, tap the Kindle Owners' Lending Library link. Your Kindle Fire displays the Browse Category pop-up menu.
5. Tap the category you want—for example, Fiction or History. Your Kindle Fire displays a list of books in that category. You can then browse the books as usual. For example, tap a book to display its details. You can then tap the Borrow For Free button (see Figure 3-1) to borrow the book.

 The Kindle Owners' Lending Library is a great way to test-drive books without buying them. If you like a book enough that you want a permanent digital copy of it, you can buy the book.

Open Library

Open Library (http://openlibrary.org/) has more than 1 million free e-books available (including different editions of the same works). You can download books in various formats, including EPUB and MOBI, but Open Library also has a Send To Kindle

FIGURE 3-1 Tap the Borrow For Free button on a book's page to borrow the book from the Kindle Owners' Lending Library as part of your Amazon Prime subscription.

feature that you can use to send a book directly from the website to your Kindle account. Your Kindle Fire then picks up the book automatically.

Project Gutenberg

Project Gutenberg (www.gutenberg.org) claims to be the first producer of free e-books and offers 38,000 e-books at this writing. All are free, but you can donate to help with the cost of running the organization.

Many of these books are in Kindle format, so you can easily read them on your Kindle Fire. Other books are only in formats such as EPUB, plain text, or HTML, which you will need to convert to Kindle format before you can read them using your Kindle Fire.

 If you find that Kindle-format e-books you download from Project Gutenberg appear in the Docs app rather than the Books app on your Kindle Fire, see Project 19 for a fix.

Normally, it's best to download books from Project Gutenberg using the browser on your computer, and then transfer the files to your Kindle Fire. If you download the books using your Kindle Fire, you will need to move the book files from your Download folder to your Books folder before the Books app will list them. You may also need to convert the books again to get them working correctly on your Kindle Fire.

Google Books

Google Books includes a fair number of free e-books. You'll need to have a Google account to download them—but you can set one up for free.

To find free e-books on Google Books, follow these steps:

1. Take your browser to http://books.google.com/ebooks.
2. Scroll down to the bottom of the page, where you'll find the Best Of The Free section.
3. Click the More link to display a page containing more free books. You can then use the links to navigate from page to page.

ManyBooks.net

ManyBooks.net (www.manybooks.net) has a wide range of e-books, all of which are free. There's a lot of overlap with Project Gutenberg, from which many of the books have come. You can browse by authors, titles, genres, languages, new titles, and recommended titles, and choose from an impressive range of different download formats.

ManyBooks.net provides a Kindle (.azw) format file for most books, but for reading on your Kindle Fire, you'll normally want to download the Mobipocket (.mobi) file.

Project 16: Add Your Existing E-book Collection to the Kindle Fire

If you already have e-books on your computer, you'll probably want to add them to your Kindle Fire so you can read them on it. This project shows you the main ways of adding e-books to your Kindle Fire.

 You may need to convert your e-books to the MOBI format before putting them on your Kindle Fire. See Project 18 for details.

Add Kindle Books You've Bought from Amazon

If you've already bought Kindle books from Amazon, you can add them to your Kindle Fire by sending them from the Amazon website. Follow these steps:

1. On your computer, open your web browser and go to the Amazon website.
2. In the Shop All Departments box on the left side, click the Kindle item to display the Kindle submenu, and then click the Manage Your Kindle link.
3. If Amazon prompts you to sign in, do so.
4. The Manage Your Kindle webpage then appears, showing your Kindle Library.

5. Click the Actions button for the book you want to put on your Kindle Fire, as shown here.

6. Click the Deliver To My link to display the Deliver Title dialog box (shown here).

7. In the Deliver To pop-up menu, choose your Kindle Fire.
8. Click the Deliver button. Amazon sends the e-book to your Kindle Fire.

Add E-books Stored in Folders

If your e-books are stored in folders, you can quickly add them to your Kindle Fire by using Windows Explorer (on Windows) or the Finder (on the Mac). Follow these steps:

1. Connect your Kindle Fire to your computer. If your Kindle Fire doesn't display the You Can Now Transfer Files From Your Computer To Kindle screen, tap the Home button to display it.
2. Open a Windows Explorer window or a Finder window showing the Kindle Fire's contents. For example:
   - **Windows**   Click the Start button, click Computer (or My Computer on Windows XP), and then double-click the Kindle drive.
   - **Mac**   Click the Finder button on the Dock to open a Finder window, and then click the Kindle drive in the Devices list in the sidebar.
3. Double-click the Books folder on the Kindle Fire to open it.
4. Open another Windows Explorer window or a Finder window to the folder that contains the e-book files you want to put on the Kindle Fire.
5. Arrange the windows so that you can see both folders.
6. Select the files on your computer, and then drag them to the Books folder.

Add E-books from Your iTunes Library

If you have e-books in the ePub format stored in your iTunes library, you can easily put them on your Kindle Fire. Follow these steps:

1. Follow Steps 1–3 of the previous list to connect your Kindle Fire to your computer and open a Windows Explorer window or a Finder window to the Books folder.
2. Launch iTunes if it's not already running.
3. Click Books in the Library category at the top of the sidebar on the left to display the list of books.
4. Arrange the windows so that you can see both iTunes and the Books folder on the Kindle Fire.
5. Select the books in iTunes, and then drag them to the Books folder (see Figure 3-2). Your computer copies the books to your Kindle Fire.

 If you need to convert a book stored in iTunes, use the Show In Explorer command or the Show In Finder command to see which folder contains the file. Right-click (or CTRL-click on the Mac) the book you want, and then click Show In Explorer (on Windows) or Show In Finder (on the Mac). iTunes opens a Windows Explorer window or a Finder window to the folder containing the book file and selects the file.

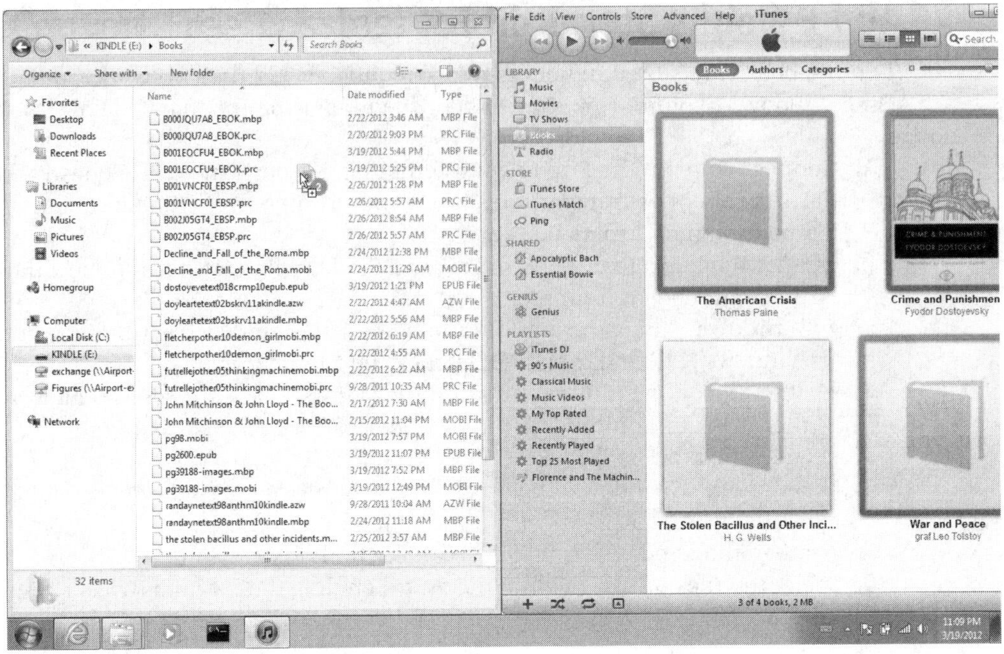

FIGURE 3-2 You can drag books from your Books collection in iTunes to the Books folder on your Kindle Fire.

Project 17: Download and Install calibre and Meet Its Interface

As you know from Project 14, if not from your own experience, e-books come in a wide variety of formats, and your Kindle Fire can handle only some of them. If you have e-books in formats your Kindle Fire can't handle, you'll need to convert the e-books to another format before you can read them. And even if the books are in the right format, you may need to reconvert them to sort out other problems (see Project 19). You may also want to create your own e-books (see Project 20).

The best tool for converting e-books and creating your own e-books is calibre, which you can download for free from the calibre website (http://calibre-ebook.com). calibre is donationware—you can donate to help keep the project funded.

Download and Install calibre

In this section, I'll show you how to download and install calibre. The process is different on Windows than on the Mac, so we'll deal with each OS separately. But the setup process is the same for each version, so we'll cover setup in a single section for both Windows and the Mac.

Download, Install, and Launch calibre on Windows

To download and install calibre on Windows, follow these steps:

1. Take your browser to the Download For Windows page, http://calibre-ebook.com/download_windows. This example uses Internet Explorer.

2. Click the link to download the Windows file to your PC. Internet Explorer displays the File Download – Security Warning dialog box (shown here). Click the Save button, choose the location in the Save dialog box, and then click the Save button.

3. When the download finishes, click the Run button in the Download Complete dialog box. Internet Explorer displays the Internet Explorer – Security Warning dialog box (shown here).

4. Click the Run button. The calibre Setup wizard runs and displays the License Agreement screen (shown here).

5. Read the agreement, select the I Accept The Terms In The License Agreement check box, and then click the Install button.

 If you want to choose the folder in which to install calibre, click the Advanced button instead of the Install button on the License Agreement screen, and then follow through the next screens. You can also choose whether to create a desktop shortcut for calibre.

6. If the User Access Control appears, click the Continue button. The calibre Setup wizard installs calibre.

7. When the Completed The calibre Setup Wizard screen (shown here) appears, make sure the Launch calibre check box is selected, and then click the Finish button. The calibre Setup wizard closes and launches calibre.

![calibre Setup window showing "Completed the calibre Setup Wizard" with "Click the Finish button to exit the Setup Wizard." and a checked "Launch calibre" box, with Back, Finish, and Cancel buttons]

Download, Install, and Launch calibre on the Mac

To download and install calibre on the Mac, follow these steps:

1. Take your browser to the Download For OS X page, http://calibre-ebook.com/download_osx.

2. Click the link to download the OS X Intel file to your Mac.

3. If OS X doesn't automatically open the disk image file for you and display a window showing its contents (as shown next), click the Downloads icon on

the Dock to display the Downloads stack, and then click the calibre disk image (.dmg) file.

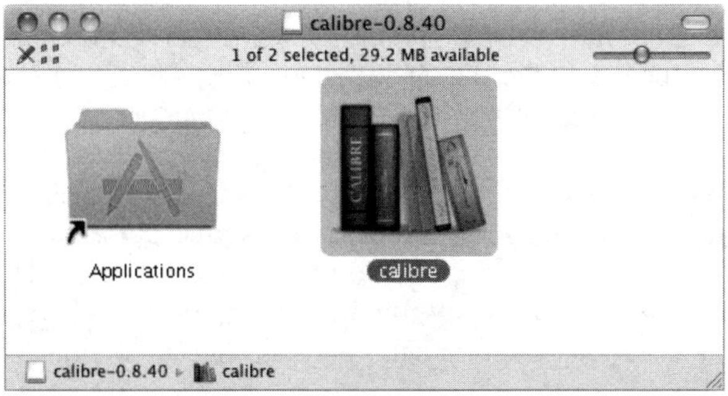

4. Drag the calibre icon onto the Applications icon. The Finder copies the calibre app to the Applications folder and displays the contents of the Applications folder.
5. Double-click the calibre icon to launch the app.

Set Up calibre

Setup works the same way for calibre on both Windows and the Mac, so we'll cover both in this section.

The first time you run calibre, the application runs the calibre welcome wizard. On the first screen (Figure 3-3 shows the Mac version), pick your language in the Choose Your Language drop-down list, and make sure the calibre library folder

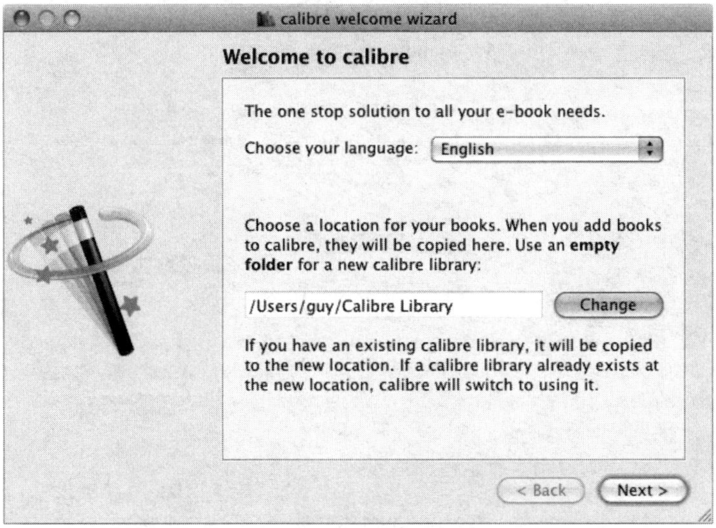

FIGURE 3-3   On the first screen of the calibre welcome wizard, choose your language and specify the folder to put your calibre library in.

suggested in the Choose A Location For Your Books box is suitable; if not, click the Change button and pick another folder.

 Use an empty folder for your calibre library rather than a folder that already contains e-books. Leave the management of this folder to calibre rather than creating and manipulating files in the folder yourself.

Click the Next button to move along to the second screen of the calibre welcome wizard (Figure 3-4 shows the Mac version). Click the Amazon item in the Manufacturers box, and then click the Kindle Fire item in the Devices box.

Click the Next button to display the final screen of the calibre welcome wizard, which congratulates you on having set up calibre. Click the Finish button to close the wizard.

Meet the calibre Interface

Next, you see the calibre window (see Figure 3-5), which has these main features:

- **Tag Browser** The Tag Browser is a pane on the left side of the calibre window that you can use to browse the e-books in your calibre Library by authors, languages, formats, publishers, and other tag information. You can toggle the display of the Tag Browser by clicking the Show/Hide Tag Browser button near the lower-right corner of the calibre window.
- **Book List** The Book List is the main part of the calibre window and shows the list of books in your calibre Library. You can sort the books in the list in different

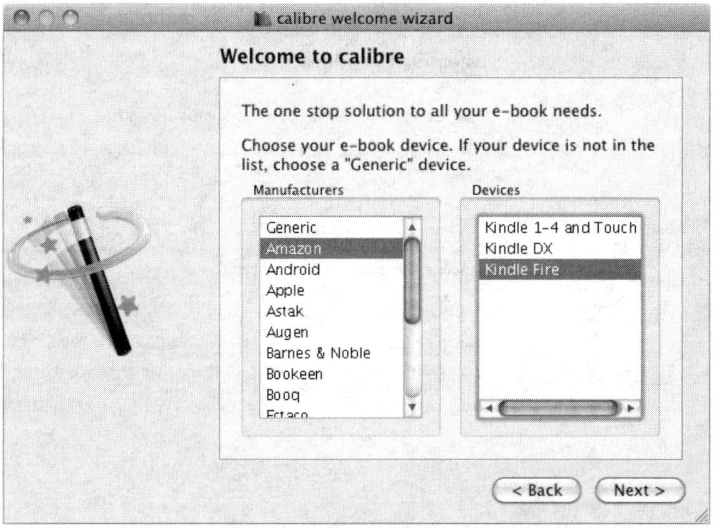

FIGURE 3-4 On the second screen of the calibre welcome wizard, choose Amazon in the Manufacturers box and Kindle Fire in the Devices box.

Book List

Book Details pane

Toolbar

Tag Browser

FIGURE 3-5   calibre's user interface puts most of the commands on the toolbar.

Show/Hide Tag
Browser button

Show/Hide Cover
Browser button

Show/Hide Book
Details Pane button

ways by clicking the column heading. For example, click the Title column heading once to sort alphabetically by title; click again to sort in reverse alphabetical order. And you can search through the books by using the Search box.

- **Book Details pane**   The Book Details pane displays details about the book selected in the Book List. You can toggle the display of the Book Details pane by clicking the Show/Hide Book Details Pane button near the lower-right corner of the calibre window.

- **Toolbar**   The toolbar contains buttons that give access to most of the commands you use in calibre. You can click the main part of each tool button to execute the command whose name appears there, or click the drop-down button at the right side of the tool button to display a menu of related commands, as shown here.

Add Your Existing E-books to Your calibre Library

Now that you've got calibre up and running, your next move is to add your existing e-books to your calibre Library so that you can use calibre to manage them, convert them, and load them on your Kindle Fire.

To add your existing e-books to your calibre Library, follow these steps:

1. Click the Add Books button to display the Select Books dialog box. This is a standard Open dialog box given a different name.

 If you need to add e-books from a folder and its subfolders, click the Add Books drop-down button, and then click the command called "Add Books From Directories, Including Subdirectories (Multiple Books Per Directory, Assumes Every Ebook File Is A Different Book)".

2. Select the e-book file or files you want to add. You can add multiple contiguous files in a single move by clicking the first file and then clicking the last file, or add individual files to your selection by CTRL-clicking them (on Windows) or ⌘-clicking them (on the Mac).
3. Click the Open button. calibre closes the Select Books dialog box and adds the e-books to your calibre Library. The book titles appear in the Book List.

Browse the E-books in Your calibre Library

You can browse by clicking the column headings in the Book List. For example:

- Click the Title column heading once to sort the e-books by title in alphabetical order. Click again to sort by title in reverse alphabetical order.
- Similarly, click the Authors column heading once or twice to sort alphabetically or reverse-alphabetically by author.
- Click the Date column heading to sort by the date you added the books. For example, you may want to identify the books you've added recently.

To quickly find the books you want, use the Cover Browser (see Figure 3-6). Click the Show/Hide Cover Browser button in the lower-right corner of the calibre window to toggle the Cover Browser on and off.

Tell calibre Not to Include the [PDOC] Tag When Creating MOBI Files

Before you create any MOBI files for your Kindle Fire, you need to set calibre not to include the [PDOC] tag in MOBI files. This [PDOC] tag tells the Kindle Fire that the MOBI file is a personal document rather than an e-book. This makes the Kindle Fire list the e-book file in the Docs app rather than the Books app, which is where you normally want your e-books to appear.

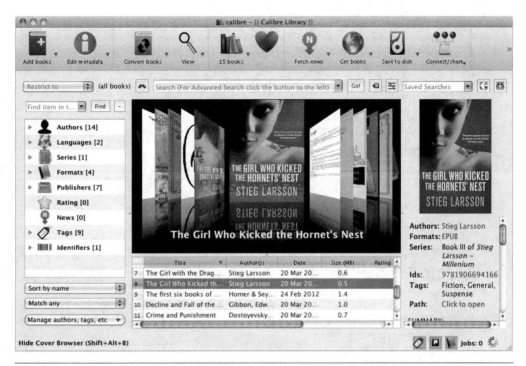

FIGURE 3-6   Use the Cover Browser to browse quickly through your books by clicking their covers.

To tell calibre not to include the [PDOC] tag when creating MOBI files, follow these steps:

1. In calibre, click the Preferences button on the toolbar to display the Preferences dialog box. Figure 3-7 shows the Preferences dialog box in Windows; the Preferences dialog box for the Mac is almost the same.
2. In the left pane, click the MOBI Output category to display its options. Figure 3-8 shows the MOBI Output category on the Output Options screen in the Preferences dialog box for calibre on Windows. The Mac version of the Preferences dialog box works in the same way.
3. Delete the [PDOC] tag from the Personal Doc Tag field.
4. Click the Apply button. calibre returns you to the Preferences dialog box.
5. Click the Close button to close the Preferences dialog box.

 If you don't change calibre's preferences so that it doesn't omit the [PDOC] tag, you can remove the tag manually on each conversion. Normally, you won't want to do this. Leaving the tag in place also means that if you use calibre's Send To Device command to convert e-books and load them straight onto your Kindle Fire, the books will include the [PDOC] tag and will appear in the Docs app rather than the Books app.

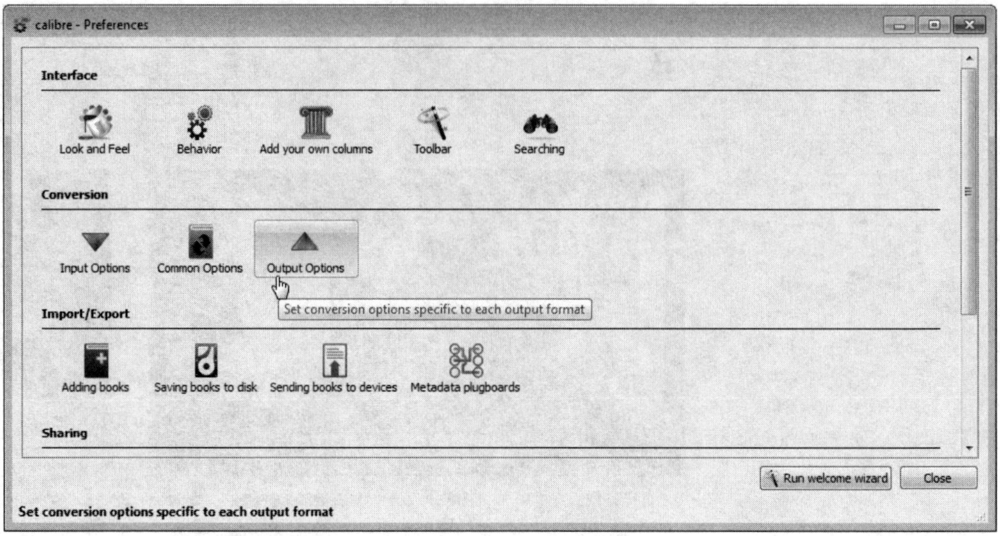

FIGURE 3-7 Click the Output Options icon in calibre's Preferences dialog box to display the Output Options screen. In the Conversion area, click the Output Options icon to display the Output Options screen.

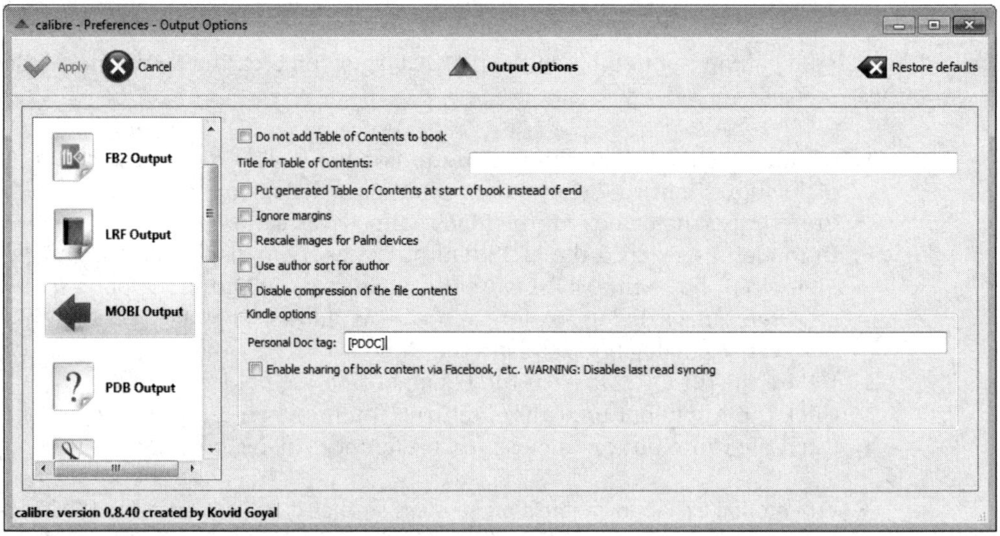

FIGURE 3-8 In the MOBI Output category on the Output Options screen in calibre's Preferences dialog box, delete the [PDOC] tag from the Personal Doc Tag field.

Connect Your Kindle Fire and Make Sure calibre Recognizes It

Now connect your Kindle Fire to your computer and make sure calibre recognizes it. Follow these steps:

1. On your computer, launch calibre if it's not already running:
   - **Windows** Choose Start | All Programs | calibre E-Book Management | calibre E-Book Management. If you have a calibre icon on your desktop, double-click that icon.
   - **Mac** Click the Launchpad icon on the Dock to display the Launchpad screen, and then click the calibre icon. If you have a calibre icon on your Dock, click that icon.
2. Connect your Kindle Fire to your computer. If your Kindle Fire doesn't display the You Can Now Transfer Files From Your Computer To Kindle screen, tap the Home button to display it.
3. Make sure the Send To Device button and Device button appear on the toolbar in calibre, as shown here. If so, calibre has recognized your Kindle Fire. If these buttons don't appear on the toolbar, calibre hasn't recognized the Kindle Fire. Try disconnecting your Kindle Fire and then connecting it again.

Transfer E-books to Your Kindle Fire from calibre

Once your calibre Library contains e-books in suitable formats, you can quickly transfer them to your Kindle Fire.

 If you need to convert an e-book to a different format before putting it on your Kindle Fire, see the next project.

To transfer an e-book from your calibre Library to your Kindle Fire, follow these steps:

1. With calibre running, connect your Kindle Fire to your computer as discussed in the previous section.
2. In the Book List, click the book you want to send to the Kindle Fire. Make sure the Book Details pane shows a format that'll work on the Kindle Fire. MOBI is best, but PRC and AZW are fine, too.

If the Book Details pane is hidden, click the Show/Hide Book Details Pane button at the lower-right corner of the calibre window to display it.

3. Click the Send To Device button on the toolbar. calibre transfers the book to your Kindle Fire.

Project 18: Convert ePub E-books to the Kindle-Friendly MOBI Format

Many sites and publishers distribute e-books in the ePub format, which the Kindle Fire can't display. To use your ePub e-books on your Kindle Fire, you need to convert them to one of the formats the Kindle Fire can use. Usually, the easiest format to use is MOBI, which is one of the Kindle Fire's native formats.

This project shows you two ways to convert e-books in the ePub format to Kindle-friendly formats. We'll start with calibre, the e-book app I showed you how to install in the previous project, and then look quickly at how you can perform the conversion using an online converter.

 You can also convert ePub files to MOBI files by using a dedicated tool such as ePub To Kindle Converter ($29.99; www.pdf-epub-converter.com/epub-to-kindle-converter.html). But given that calibre is free, easy to use, and performs the conversion well, calibre is normally a better choice.

Convert ePub E-books to MOBI Format Using calibre

If you've installed it, calibre is the best tool for creating MOBI files from ePub files. To convert an ePub e-book, follow these steps:

1. Add the ePub file to your calibre Library as explained in Project 17, earlier in this chapter.
2. Click the book in the Book List to select it.
3. Click the Convert Books button on the toolbar to display the Convert dialog box. At first, the Convert dialog box displays the Metadata category (see Figure 3-9).
4. In the Output Format drop-down list in the upper-right corner, choose the MOBI item.
5. In the other fields in the upper-right corner, add any book information that's missing. For example, if the Author box is empty, type the author's name. You can also add tags in the Tags box—for example, Crime, History, or Textbook—to help you sort your books more easily.
6. If the book is missing a cover image, or if you want to change the existing cover image, click the button at the right end of the Change Cover Image box. In the Choose Cover For dialog box that opens, select the cover image to use, and then click the Open button.

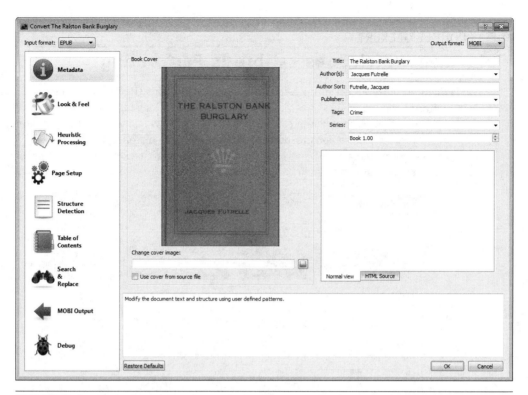

FIGURE 3-9   In the Metadata category in the Convert dialog box, choose MOBI in the Output Format drop-down list, make sure the book information is correct, and add a cover image if necessary.

 At this point, if you haven't removed the [PDOC] tag in calibre's settings, you can remove it manually. In the Category box on the left, click the MOBI Output category to display the MOBI Output options. In the Personal Doc Tag box, select the [PDOC] tag, and then press DELETE to delete it.

7. Click the OK button to send the conversion job to the jobs queue.

Convert ePub E-books and Transfer the MOBI Files to Your Kindle Fire

Converting ePub files to MOBI files is easy enough, as you saw in the previous section. But you can get ePub files onto your Kindle Fire even more smoothly by using calibre's feature for converting the file and immediately transferring it to the Kindle Fire.

DOUBLE GEEKERY

See Which Jobs calibre Is Performing

When you take an action in calibre, the application adds the job to its Jobs list and performs the job when it has finished performing previous jobs. The Jobs readout in the lower-right corner of the calibre window shows how many jobs calibre is working on. You can click the Jobs readout to display the Jobs dialog box (shown here), which shows your current and past jobs. Here, you can check on the status of jobs and stop any jobs that seem to have gotten stuck.

Click the Close button (the × button on Windows, the red button on the Mac) when you finish using the Jobs dialog box.

To convert an ePub e-book to a MOBI file and put that file on your Kindle Fire, follow these steps:

1. On your computer, launch calibre if it's not already running:
   - **Windows**   Choose Start | All Programs | calibre E-Book Management | calibre E-Book Management. If you have a calibre icon on your desktop, double-click that icon.
   - **Mac**   Click the Launchpad icon on the Dock to display the Launchpad screen, and then click the calibre icon. If you have a calibre icon on your Dock, click that icon.
2. Connect your Kindle Fire to your computer. If your Kindle Fire doesn't display the You Can Now Transfer Files From Your Computer To Kindle screen, tap the Home button to display it.
3. Make sure the Send To Device button appears on the toolbar in calibre. If so, calibre has recognized your Kindle Fire. If the Device button doesn't appear on the toolbar, calibre hasn't recognized the Kindle Fire. Try disconnecting your Kindle Fire and then connecting it again.

4. In the Book List, click the e-book you want to convert to MOBI format and send to the Kindle Fire.
5. On the toolbar, click the Send To Device drop-down button, click the Send Specific Format To submenu, and then click the Main Memory item on the submenu. calibre displays the Choose Format dialog box (shown here).

6. In the Choose Format To Send To Device list, click the MOBI item.
7. Click the OK button. The first time you do this, calibre displays the No Suitable Formats dialog box, shown here, asking if you want to automatically convert the book before uploading it to the Kindle Fire.
8. Clear the Show This Confirmation Again check box to prevent calibre from displaying this dialog box in the future.
9. Click the Yes button to tell calibre to perform the conversion. calibre then converts the book and puts it on your Kindle Fire.

Convert ePub E-books to Kindle-Friendly Formats Using an Online Converter

If you don't have calibre, you can convert ePub files to MOBI files by using an online converter such as the one at Convert Files (www.convertfiles.com), shown in Figure 3-10. You select your input file, specify which format it's in and which output format you want (.mobi), and then click the Convert button. After converting the file, the converter gives you a link to a page to download the MOBI file.

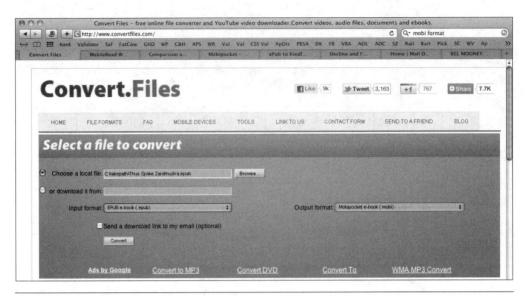

FIGURE 3-10 You can quickly convert an ePub e-book to a MOBI file by using an online converter such as the one at Convert Files.

Project 19: Make Your E-books Appear in the Books App Rather Than the Docs App

As you saw in the previous project, you need to delete the default [PDOC] tag when converting an ePub file to the MOBI format. Otherwise, you add the e-book to your Kindle Fire marked as a personal document, which makes it appear in the Docs app rather than in the Books app.

 Having an e-book appear in the Docs app rather than the Books app isn't the end of the world. When you tap the e-book in the Docs app, it opens for reading in just the same way as when you tap it in the Books app. But you'll probably want to keep your e-books separate from your documents so that you can easily find the books you want. And the Books app lets you sort the books by author as well as by title and by recent books, whereas the Docs app can sort only by title and by recent documents.

In fact, it's not just MOBI files that can have this problem. The Kindle Fire lists *any* e-book that contains the [PDOC] tag in the Docs app rather than the Books app. So if you load an e-book in any of the formats the Kindle Fire can display on the Kindle Fire and it appears not in the Books app but in the Docs app, it may contain the [PDOC] tag.

If an e-book you've added to your Kindle Fire appears in neither the Books app nor the Docs app, the e-book is either in a format the Kindle Fire can't handle or is corrupt. Try opening the e-book in calibre's E-Book Viewer to see if the file is okay: In calibre, double-click the e-book in the Book List. If the e-book looks okay, export it as a MOBI file and put it on your Kindle Fire again. If not, you'll need to download the file again from wherever you got it.

To solve this problem, you must remove the [PDOC] tag from the e-book file. calibre doesn't provide a command for removing the tag from an existing file, so you need to convert the e-book file to a MOBI file again, this time removing the tag.

To convert the e-book and remove the tag, follow these steps:

1. Add the ePub file to your calibre library as explained in Project 17, earlier in this chapter.
2. Click the book in the Book List to select it.
3. Click the Convert Books button on the toolbar to display the Convert dialog box. At first, the Convert dialog box displays the Metadata category (shown in Figure 3-9, earlier in this chapter).
4. In the Output Format drop-down list in the upper-right corner, choose the MOBI item.
5. If you haven't deleted the [PDOC] tag from the Personal Doc Tag field in the MOBI Output category on the Output Options screen in the Preferences dialog box for calibre, you need to remove the tag manually. Click the Category box on the left, click the MOBI Output category, and then delete the [PDOC] tag from the Personal Doc Tag box.
6. Click the OK button to send the conversion job to the jobs queue. calibre creates a MOBI file of the e-book and copies the file to your Kindle Fire.

Project 20: Convert E-books, PDFs, Documents, and Webpages to Kindle Fire Format

As you've seen in the preceding several projects, calibre is great for converting e-books from various formats to MOBI files for reading on your Kindle Fire. But you may need to convert e-books when calibre isn't available—for example, when you're using someone else's computer. Or you may want to create e-book files out of PDF files, documents, or webpages so that you can read them easily on your Kindle Fire anywhere.

To convert files quickly and easily, you can use Amazon's Kindle Personal Document Service. This is an automatic service to which you send files via e-mail using your Kindle address. The service converts the files to Kindle-friendly formats and makes them available to your Kindle Fire, which picks them up at the next sync.

You can find the address to use for the Kindle Personal Document Service in either of these ways:

- **On your Kindle Fire** Open the Docs app and look at the Send Documents To readout at the top.
- **On your computer** Follow these steps:
    1. Open your web browser and go to Amazon.com.
    2. In the box on the left, click the Kindle link, and then click the Manage Your Kindle link on the submenu.
    3. If Amazon.com displays the login screen, log in as usual.
    4. In the Your Kindle Account box on the left, click the Personal Document Settings link.
    5. On the Personal Document Settings page, look at the Send-To-Kindle E-Mail Settings area.

On the Personal Document Settings page in the Manage Your Kindle area on Amazon.com, you can also manage the e-mail addresses allowed to send you content for adding to your Kindle Fire. Click the Add A New Approved E-Mail Address link to start adding a new address. For example, you may need to send yourself documents from multiple addresses, or you may choose to let specific other people send you items to read.

Once you know the address to use, create a message in your e-mail application and put that address in the To field. Attach the e-book files or other files you want to put on your Kindle Fire, and then send the message.

If the Kindle Personal Document Service can't convert any of the files you send, the service sends you an e-mail message telling you which file it had a problem with.

DOUBLE GEEKERY

Which File Formats Can the Kindle Personal Document Service Convert?

The Kindle Personal Document Service can convert these file formats to Kindle format:
- MOBI books (.mobi files)
- Microsoft Word documents (.doc files or .docx files)
- Plain-text documents (.txt files)
- Rich-text format documents (.rtf files)
- Webpages and HTML documents (.htm files and .html files)
- Zip files (.zip files)
- X-Zip files (.x-zip files)

You can send up to 25 documents in a single e-mail message, which is enough for most people.

Project 21: Read Your E-books in the Bath—Safely

Your Kindle Fire is brilliant for reading—but as an electronic device, it's emphatically not designed for reading in the bath. Just one slip of the hand and odds are you're out $200.

 This project assumes you want to read in the bath, but there's nothing to stop you using your Kindle Fire as a shower radio, beach movie player, or other tool for adding media to your active lifestyle. Music and other audio are usually adequately audible through a waterproof case, but the case certainly doesn't improve the sound.

If you want to read in the bath, or in other wet or dirty conditions, you need to protect your Kindle Fire. That means using a waterproof case or simply sealing the Kindle Fire in a heavy-duty plastic bag.

You can find a fair variety of waterproof cases in online stores. There are three main types:

- **Skins**   These are soft, tight-fitting covers made of thin plastic or polymer. Many skins are designed for only a single use, so they're good for a day at the beach but not for reading in the bath every night. For an example of a Kindle Fire skin, see the iOttie ($19.99 for a two-pack; www.iottie.com).

 Fully waterproof cases are great if that's what you need, but because they're sealed, they tend to make access to the Kindle Fire's ports difficult. You can use the screen as usual through the case, which is great for reading, but you'll typically need to remove all or part of the case in order to recharge the Kindle Fire or to connect headphones. The bigger waterproof cases simply snap open, but those that fit more snugly can take time and effort to remove.

- **Kindle Fire–specific cases**   Some waterproof cases are built to fit only the Kindle Fire. These cases tend to be more expensive, but they're designed for long-term use, and you can remove them so that you can use headphones or charge or load the Kindle Fire.
- **Waterproof pockets**   Many waterproof cases are, in effect, thick plastic bags big enough for the Kindle Fire or similar-size devices, with a hermetic seal applied. Some have fancier designs than others and carry higher prices, but even the basic models can be effective if you use them carefully. Figure 3-11 shows an example of the basic waterproof case I use. It has three press-to-seal strips at the top, which folds over and then buttons down.

DOUBLE GEEKERY

Know What the IPX Certifications Mean

When you're shopping for waterproof cases, you'll see certification numbers such as IPX7 and IPX8. Here, IP stands for *ingress protection*—how much protection the case provides against stuff getting in. The following list shows what the IPX ratings mean for liquid ingress protection.

IPX Rating	Protection Against
1	Vertically dripping water
2	Water dripping at an angle of up to 15 degrees
3	Water spraying at an angle of up to 60 degrees
4	Water splashing from any direction
5	Water jets from any direction
6	Powerful water jets
7	Immersion up to 1 meter deep
8	Immersion of more than 1 meter deep

So if you want your Kindle Fire to survive a drop into the bath, IPX7 has you covered. If you're planning to take your Kindle Fire on a boat trip that involves a chance of falling overboard, you'll want IPX8, which typically means the case is hermetically sealed.

FIGURE 3-11 A waterproof case lets you read books in the bath without worrying about damaging your Kindle Fire.

Even a teaspoon of water in the wrong place can ruin your Kindle Fire, so you'll want to be sure you can trust the case you buy. This is one area where saving money by buying a no-name brand can cost you dear, so you may decide that you want to stick with a big-name brand—perhaps one that provides a guarantee. Alternatively, test the case by putting a kitchen towel rather than your Kindle Fire in it and then immersing it. You'll quickly learn whether the case is truly waterproof.

Because the Kindle Fire is an Amazon product, Amazon is arguably the best place to find a waterproof case, because you can read the buyer reviews to get a clearer idea of what a particular case is good for, how well it delivers on its promises, and what its weak points are. But you'll find plenty of waterproof cases on eBay as well.

 DOUBLE GEEKERY

What to Do When Your Kindle Fire Gets Wet

If your Kindle Fire gets wet, don't panic—it's not the end of the world, and it's not necessarily the end of your Kindle Fire.

Some Kindle Fires recover after being dunked in water. Others don't. It all depends on how far the water gets in and what damage it causes. But you should try drying your Kindle Fire out thoroughly in case the water hasn't wrecked it.

So if the Kindle Fire is still on, turn it off completely. Press and hold the power button until the Do You Want To Shut Down Your Kindle? screen appears, and then tap the Shut Down button.

Next, put the Kindle Fire to dry somewhere with gentle warmth that you can leave it for two or three days if necessary. Make sure you don't bake the Kindle Fire: Don't put it right on top of a radiator or leave it in direct sunlight.

When you've put the Kindle Fire to dry, round up some desiccant gel—you know, those little bags marked "Do Not Eat" that come in the packaging of electronics equipment. Chances are, you can get these from your local hardware store; if not, you'll find plenty online. Try Amazon first, and then see if you can find a better price on eBay.

When you get the desiccant gel, line a box with the bags, put the Kindle Fire in it, and then put more of the bags on top. Then put the box back in that warm, comfortable spot.

If you want to get the drying started more quickly, you can use uncooked rice instead of desiccant gel—but make sure you don't get the rice in the Kindle Fire's ports. To use rice, fill a thin sock or a section of pantyhose with rice, and then wrap the Kindle Fire in it. The rice will gradually absorb the moisture, drying out the Kindle Fire.

After drying out for two or three days, try powering the Kindle Fire on. If you get no reaction rather than the startup logo or electronic spasms of death, the battery may be out of power. So connect your Kindle Fire to a power supply and try again.

Project 22: Read Your Kindle Books on Your Computer—or Someone Else's

Even if you take your Kindle Fire everywhere with you, at times you'll probably need to read on a larger screen. With Amazon's Kindle Reader application, you can read your Kindle books on any PC or Mac. And with Kindle Cloud Reader, you can read your Kindle books on any computer or device that has a web browser.

Read Books on Your PC or Mac with the Kindle Application

To read your Kindle books on your PC or Mac, download and install the Kindle Reader application for your operating system. Follow these general steps:

1. Open your web browser and go to www.amazon.com.
2. In the Shop All Departments box, click the Kindle button, and then click Free Kindle Reading Apps on the submenu. You'll see the Free Kindle Reading Apps screen.
3. Click the Windows PC button or the Mac button, as appropriate. The Windows PC pane or the Mac pane appears.
4. Click the Download Now button to download the file.
5. Install the Kindle For PC application or the Kindle For Mac application as usual:
   - **Windows**   When the dialog box shown here appears, click the Run button. In the Open File – Security Warning dialog box, click the Run button. Then enter your account details in the Register Kindle dialog box and click the Register button. The Kindle For PC application then runs automatically.

   - **Mac**   When the download completes, open the Kindle For Mac disk image file if your Mac doesn't open it for you. For example, click the Downloads button on the Dock, and then click the KindleForMac disk image file. In the Finder window that opens, drag the Kindle.app icon onto the Applications icon. Double-click the Applications icon to display the Applications folder, and then double-click the Kindle icon. Enter your account details in the Register Kindle dialog box, and then click the Register button.

Once you get the Kindle application up and running, your library appears, as shown in Figure 3-12. You can then click a book to open it for reading.

FIGURE 3-12   After opening Kindle For PC or Kindle For Mac (shown here), click the cover of the book you want to read.

Read Books in Any Web Browser with Kindle Cloud Reader

As you saw earlier in this project, you can install Kindle Reader on any PC or Mac in a few minutes. So Kindle Reader will likely be your preferred tool for reading your Kindle books on your own computers.

 DOUBLE GEEKERY

Get a Web Browser That Supports Kindle Cloud Reader

To use Kindle Cloud Reader, your computer or device must have a suitable browser. These are the browsers that work at this writing:

- **Windows**   Firefox 6 or later, Chrome 11 or later, or Safari 5 or later
- **Mac**   Safari 5 or later, Firefox 6 or later, or Chrome 11 or later
- **iPad**   Safari

If you're using Windows, you'll need to install Firefox, Chrome, or Safari if you don't have one of these browsers. Internet Explorer won't work.

If you have a Mac or an iPad, you're all set, because Safari comes built into OS X and iOS.

But when you're using someone else's computer and the owner doesn't want you to install Kindle Reader or actively prevents you from doing so, you can use Kindle Cloud Reader. Follow these steps:

1. Open Firefox, Safari, or Chrome. I'll use Safari on the Mac in this example.
2. Go to the Amazon.com website.
3. In the Shop All Departments box, click the Kindle button, and then click Kindle Cloud Reader on the submenu. You'll see the Kindle Cloud Reader screen (see Figure 3-13).
4. Click the Sign In Or Create New Account button.
5. If your browser displays the Sign In screen, type your e-mail address and password, and then click the Sign In Using Our Secure Server button.
6. In the Set Up Kindle Cloud Reader For Offline Reading dialog box that appears (shown here), click the Enable Offline button if you want to be able to read your Kindle books when this computer is offline. You then follow through a simple

FIGURE 3-13   On the Kindle Cloud Reader screen, click the Sign In Or Create New Account button to get started.

DOUBLE GEEKERY

Turn on JavaScript in Your Browser

Kindle Cloud Reader requires your browser to be running JavaScript. If you've turned JavaScript off as a security measure, you'll need to turn it back on in order to get Kindle Cloud Reader working.

Here's how to turn JavaScript on:

- **Safari (Windows or Mac)**   On Windows, click the cog-wheel button, and then click Preferences to display the Preferences window; on the Mac, choose Safari | Preferences. Click the Security tab, and then select the Enable JavaScript check box in the Web Content area. Close the Preferences window.
- **Safari (iPad)**   From the Home screen, tap the Settings icon to display the Settings screen. Tap the Safari button in the left column to display the Safari pane, and then move the JavaScript switch in the Security area to the On position.
- **Firefox**   On Windows, choose Firefox | Options to display the Options dialog box; on the Mac, choose Firefox | Preferences to display the Preferences dialog box. Click the Content tab, select the Enable JavaScript check box, and then close the dialog box.
- **Chrome**   On Windows, click the wrench icon, and then click Options to display the Options tab; on the Mac, choose Chrome | Preferences to display the Preferences tab. On that tab, click Under The Hood in the left pane. In the Privacy area, click the Content Settings button to display the Content Settings pane. In the JavaScript section, select the Allow All Sites To Run JavaScript option button. Then click the × button to close the pane, and click the × button on the Options tab.

installation procedure for Cloud Reader. Normally, you'll want to do this only if you're using your own computer. On any other computer, click the Not Now button.

7. When you see the books in your library, as shown in Figure 3-14, you can click a book's cover to open the book for reading.

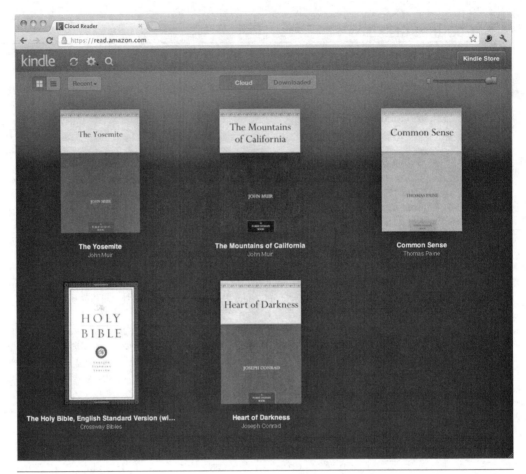

FIGURE 3-14 Click a book to start reading it in Cloud Reader.

4 System and Apps Geekery

In this chapter, we'll get your geek on by working with system items and apps. We'll start by looking at how to lock your Kindle Fire for privacy, update the tablet with the latest operating system fixes, and update the apps you use. We'll then move along to troubleshooting lockups and crashes, making your Kindle Fire (in general) and the Silk browser (in particular) run faster, and squeezing more life out of the battery.

After that, it's time for some self-preservation. I'll show you how to prevent your children from going online with your Kindle Fire, how to turn off One-Click Buy to spare your credit cards work, and how to protect your Kindle Fire against malware.

Next, you'll learn how to get apps from other sources than Amazon's Appstore and how to side-load such apps onto the tablet. We'll then examine how to make the most of the Kindle Fire's Wi-Fi capabilities before digging into secrets for entering text more quickly in documents.

We'll finish by customizing the Carousel or replacing it with a better launcher; recovering space from the Kindle Fire's storage; putting your photos on the tablet; and using it to keep up with your social networks on Facebook, LinkedIn, and Twitter.

Project 23: Lock Your Kindle Fire for Privacy

Like any useful tablet computer, your Kindle Fire probably contains plenty of private information to make it attractive to a thief or attacker. For example, your Kindle Fire may contain

- Your contacts' names and addresses—great for phishers and spammers
- Access to your e-mail account—just perfect for an attacker
- The ability to buy a vast variety of products on Amazon with your credit card—and have them delivered next day courtesy of Amazon Prime

That means it's vital you lock your Kindle Fire for privacy.

Locking is easy—but most people neglect it. Don't make this mistake.

Turn On Locking

To turn on locking for your Kindle Fire, follow these steps:

1. Tap the Home button to go to the Home screen.
2. Tap the Settings icon on the status bar to display the Settings icons.
3. Tap the More button to display the Settings screen.
4. Tap the Security button to display the Security screen (see Figure 4-1).
5. Tap the Lock Screen Password switch and move it to the On position. Your Kindle Fire displays the Lock Screen Password screen (see Figure 4-2).
6. Type your password in the Enter Password box. The password must be at least four characters long.
7. Tap in the Confirm Password box and type the password again.
8. Tap the Finish button. Your Kindle Fire displays the Security screen again.
9. Tap the Home button to return to the Home screen.

Lock Your Kindle Fire

After you turn on locking and set a password, you can lock your Kindle Fire simply by pressing the power button and putting the Kindle Fire to sleep. The Kindle Fire also locks automatically when it puts itself to sleep.

FIGURE 4-1 On the Security screen in Settings, tap the Lock Screen Password switch and move it to the On position to start securing your Kindle Fire.

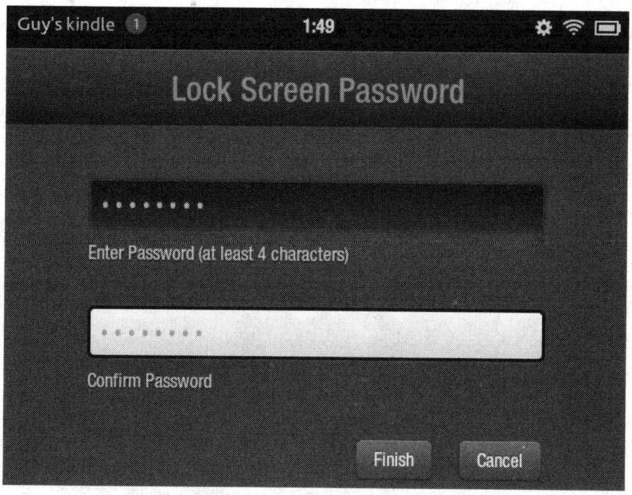

FIGURE 4-2   On the Lock Screen Password screen, type the locking password twice, and then tap the Finish button.

See the section "Set Your Kindle Fire to Go to Sleep After a Short Delay" in Project 29, later in this chapter, for instructions on getting your Kindle Fire to go to sleep quickly.

DOUBLE GEEKERY

Understand How to Create Strong Passwords

A *strong* password is one that is impossible to guess and very difficult to break with a dictionary attack, an attack that tries to find the password by supplying the words in a list (called a *dictionary*) one at a time.

To create a strong password, follow these rules:

- Use eight characters or more.
- Don't use any real word in any language.
- Include at least one number.
- Include at least one symbol (for example, ! or *).

Unlock Your Kindle Fire

To unlock your Kindle Fire, slide the slider on the lock screen as usual. The Kindle Fire then displays the Enter Password To Unlock screen (see Figure 4-3).

Type your password, and then tap the OK button in the lower-right corner of the keyboard. Your Kindle Fire then unlocks, and you can use it as normal.

 If you need to turn the lock off, go back to the Security screen in Settings, tap the Lock Screen Password switch, and move it to the Off position. Your Kindle Fire displays the Lock Screen Password screen, on which you must type the password and tap the Finish button to remove the password.

FIGURE 4-3 On the Enter Password To Unlock screen, type your password and then tap the OK button.

Project 24: Update Your Kindle Fire with the Latest Fixes

To keep your Kindle Fire running well, it's a good idea to update it with the latest fixes that Amazon releases. Your Kindle Fire automatically updates itself with new versions of its operating system when it finds them, but you may sometimes want to apply an update before your Kindle Fire finds it—for example, because you've heard that it fixes a particular problem or a dangerous security hole.

To update your Kindle Fire to the latest version of its operating system, follow these steps:

1. Tap the Home button to go to the Home screen.
2. Tap the Settings icon on the status bar to display the Settings icons.
3. Tap the More button to display the Settings screen.
4. Tap the Device button to display the Device screen (see Figure 4-4).
5. In the middle of the screen, tap the Update Your Kindle button.

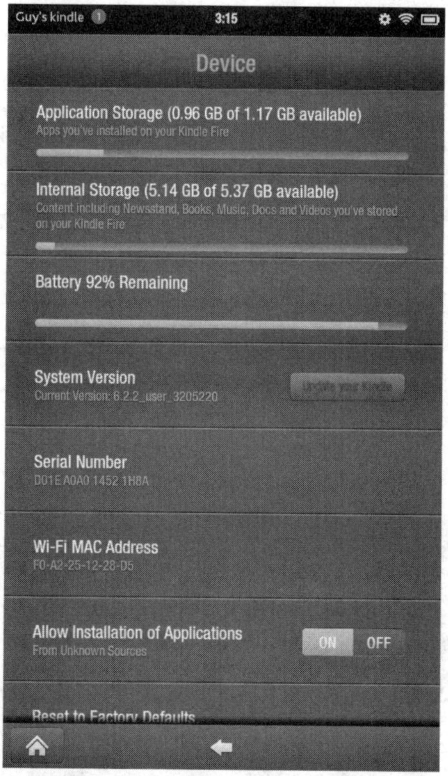

FIGURE 4-4   Tap the Update Your Kindle button on the Device screen to start getting the latest updates.

If the Update Your Kindle button is dimmed out, there's no available update for your Kindle Fire at the moment.

6. Wait while the Kindle Fire downloads the update and installs it. Most updates require the Kindle Fire to restart once or twice, so leave it alone until it finishes.

 At this writing, the Kindle Fire's user interface gives you no way to turn off automatic updates. If you need to turn off automatic updates, you must "root" the Kindle Fire, taking control of it at the system level. This book doesn't cover rooting.

Project 25: Keep Your Apps Updated

To get the best performance and most features out of your apps, make sure you keep them updated to their latest versions.

To update your apps, follow these steps:

1. Tap the Home button to display the Home screen.
2. Tap the Apps button to display the Apps screen.
3. Tap the Store button in the upper-right corner to display the Store screen.
4. Tap the Menu button at the bottom to display the Menu panel (shown here).

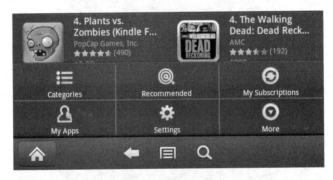

5. Tap the My Apps button to display the My Apps screen (shown here).

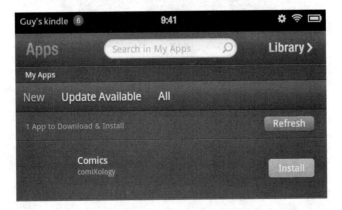

6. Tap the Update Available button to display the Update Available screen (see Figure 4-5).

7. Tap the Update button for each app you want to update. The Apps screen shows you the progress of each download and installation, as shown here.

8. When you finish updating your apps, tap the Library button to go back to the

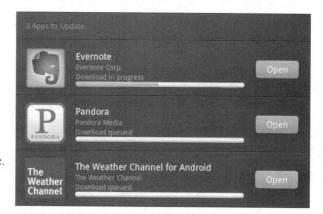

Apps screen on your Kindle Fire. You can then launch the updated apps and see what's new.

FIGURE 4-5   On the Update Available screen, tap the Update button for each app you want to update.

Project 26: Troubleshoot Lockups and Crashes

Your Kindle Fire's hardware and software are as stable as Amazon has been able to make them—but as with all hardware and software, problems can still occur.

In this project, we'll look at how to force an app to close if it stops responding and how to perform a hard reset when the Kindle Fire locks up or won't power on.

Force an App to Close

If an app stops responding, you may need to force it to close. Here's how to do that:

1. Tap the Settings button on the status bar to display the settings bar.
2. Tap the More button on the settings bar to display the Settings screen.
3. Tap the Applications button to display the Applications screen.
4. Tap the Filter By pop-up menu, and then tap Running Applications to display the list of running apps, as shown in Figure 4-6.

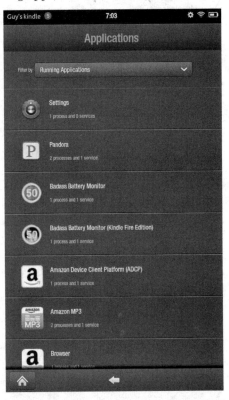

FIGURE 4-6 On the Applications screen in the Settings app, choose Running Applications in the Filter By pop-up menu to show the list of apps that are running.

FIGURE 4-7 On the screen for an app, tap the Force Stop button to force the app to stop.

5. Tap the app you want to stop. The Kindle Fire displays the app's screen. Figure 4-7 shows the screen for the Pandora app.
6. Tap the Force Stop button. The Kindle Fire displays the Force Stop dialog box (shown here).

7. Tap the OK button. The Kindle Fire forces the app to stop and then displays the app's screen again.
8. Tap the Back button to return to the Applications screen in the Settings app.

Deal with System Lockups by Performing a Hard Reset

As you saw in the previous section, you can usually deal with an app crash by forcing the app to stop. But if your Kindle Fire's Android operating system locks up, you'll usually need to perform a hard reset to get the Kindle Fire working again.

 Before you perform a hard reset, give your Kindle Fire a couple of minutes to sort itself out. Sometimes, Android can clear whatever problem has occurred behind the scenes. Your Kindle Fire then starts responding to the touch screen again.

When you're sure you need to perform a hard reset, press and hold the power button for about 20 seconds. If your Kindle Fire displays the Do You Want To Shut Down Your Kindle? screen (shown here), release the power button, and then tap the Shut Down button. If not, keep holding down the power button until the screen goes dark, then release the button.

Give your Kindle Fire a few seconds of rest before you press the power button to turn it on again.

Troubleshoot Startup Problems

If your Kindle Fire won't start when you press the power button, make sure it's not simply out of battery power. For example, connect it to the USB charger and give it a few minutes to charge.

If your Kindle Fire still won't start, try performing a hard reset, as described in the previous section. You may need to do this several times, but normally the Kindle Fire will start responding.

If your Kindle Fire still won't start up, contact Amazon Customer Service for advice.

 DOUBLE GEEKERY

Reset Your Kindle Fire to Factory Defaults

If the software on your Kindle Fire gets severely messed up, you may need to reset the Kindle Fire to its factory default settings. Doing so removes all your content, all the apps you've installed, and all the settings you've chosen, so you'll normally want to do this when all other troubleshooting has failed. In this case, you'll need to spend time preparing for the reset and time restoring your content, apps, and settings afterward.

You can reinstall your apps from the Appstore, restore your purchased items from Amazon, and restore the data you've put on your Kindle Fire from your computer by transferring it again. So you don't need to back up your entire Kindle Fire before resetting it to factory defaults. But you do need to back up any content you've created on your Kindle Fire but not yet transferred to your computer or online storage. For example, if you have created documents in Quickoffice, you will need to store them somewhere safe to make sure you don't lose them.

You may also want to perform a reset to factory defaults before giving (or selling) your Kindle Fire to someone else.

When you're ready to perform the reset to factory defaults, follow these steps:

1. Tap the Settings icon on the status bar to display the settings bar.
2. Tap the More button to display the Settings screen.
3. Scroll down to the bottom of the Settings screen.
4. Tap the Device button to display the Device screen (shown in Figure 4-4, earlier in this chapter).
5. Scroll down to the bottom of the Device screen.
6. Tap the Reset To Factory Defaults button. Your Kindle Fire displays the Factory Data Reset screen (shown here).

Factory Data Reset

You are about to reset your Kindle to factory defaults, which will remove all of your personal information, Amazon account information, downloaded content and applications.

Do you wish to Continue?

| Erase everything | Cancel |

7. Tap the Erase Everything button. Your Kindle Fire deletes your content, settings, and installed apps, and returns itself to its factory settings.

After your Kindle Fire restarts, establish a wireless network connection so that you can reinstall your apps from the Appstore and download your purchased items from Amazon again. Then connect your Kindle Fire to your computer and transfer the music, video, and other files that you keep on the tablet.

Project 27: Make Your Kindle Fire Run Faster

Your Kindle Fire focuses on delivering a great user experience via content and usability rather than by running at blinding speed. Amazon has kept the Kindle Fire's hardware relatively modest in order to keep the price down, with the result that the tablet usually runs at an acceptable speed—but you may sometimes notice a lag or find you have to tap a button more than once to get a reaction.

This lack of blazing speed is normal, and you can't do a great deal to speed up your Kindle Fire without rooting it (which this book doesn't cover). But if you find your Kindle Fire seems to be running more slowly than usual, you can bring it back up to its normal speed.

This project explains the three moves you'll normally need:

- Restarting your Kindle Fire
- Closing any running apps you're no longer using
- Updating your Kindle Fire's system software to the latest version

 If the Kindle Fire's Silk browser seems to be running slowly, see the next project for remedies you can apply.

Restart Your Kindle Fire

Your Kindle Fire has 512MB of RAM squirreled away inside its case, and it's pretty smart about managing it. Even so, if you run too many apps at once, your Kindle Fire can run low on memory. When this happens, your Kindle Fire may start responding more slowly or jerkily.

Often, the easiest and most effective way to clear any unnecessary processes out of your Kindle Fire's memory is to restart the Kindle Fire. You won't always want to restart, because it's awkward if you're in the middle of work or play. But if you have time, treat a restart as your first speedup move.

To restart the Kindle Fire, follow these steps:

1. Finish or save any unsaved work that you care about. For example, if you're composing an e-mail message, either finish it and send it, or save it as a draft so that you can finish it later.
2. Press and hold the power button until the Kindle Fire displays the Do You Want To Shut Down Your Kindle? message.
3. Tap the Shut Down button. The Kindle Fire turns itself off.
4. Press and hold the power button until the Kindle Fire logo appears on the screen; then let the Kindle Fire finish starting up.

Close Apps You're No Longer Using

If you're not prepared to restart your Kindle Fire to get it running faster, close any apps you're no longer using. If you've been using the tablet for a while, you may have lots of apps running in the background—maybe too many for Android to handle smoothly.

To avoid running into this problem, it's a good idea to close any apps you're no longer using. If you still use a desktop or laptop, you're probably used to doing this.

But on the Kindle Fire, closing apps is more difficult. This is because some apps close automatically when you leave them by returning to the Home screen, but other apps keep running until you close them explicitly.

Some apps have an explicit Exit command. The first place to look for this is on the Menu panel. For example, TuneIn Radio has an Exit button, as shown here, for closing the app.

So far, so easy—but TuneIn Radio then hits you with a confirmation dialog box, as shown here, so you need to tap the OK button to regain your freedom.

Other apps make the Exit command harder to find, so you need to jump through an extra hoop to leave the app—perhaps to stop you exiting the app by accident. For example, here's how you close the BeyondPod Podcast Manager app that I recommended in Chapter 1:

1. Tap the Menu button to display the Menu panel (shown here).

2. Tap the More button to display the More dialog box (see Figure 4-8).
3. Tap the Exit BeyondPod button.

 You may also need to force an app to close. See the section "Force an App to Close" in Project 26 for instructions.

FIGURE 4-8   Some apps include an explicit Exit command, such as the Exit BeyondPod button shown at the bottom of the More dialog box here.

DOUBLE GEEKERY

Watch Out for Buttons That Look Like Exit Buttons, But Aren't

Be careful when you're looking for Exit commands on apps, because you may run into surprises. For example, Evernote's Menu panel contains a Sign Out button that you might think would close Evernote. But what the Sign Out command does is remove all your Evernote data from your Kindle Fire—which probably isn't what you want to do. Luckily, Evernote displays a confirmation dialog box (shown here) to make clear what the command does.

> ⚠ **Sign out**
>
> Signing out will remove all your cached notes and offline notebooks from your device.
> Do you want to sign out?
>
> [Yes]   [No]

Install Software Updates When They Become Available

Since Amazon released the Kindle Fire, it has released several software updates that bring bug fixes, new features, and speed improvements. So you'll probably want to install each new software update as soon as it becomes available. Look back to Project 24 for details of how to install an update.

 Your Kindle Fire is set to download and install software updates automatically, so if you're prepared to wait for a few days, you don't need to install updates manually.

Project 28: Make the Silk Browser Load Pages More Quickly

Amazon designed the Silk browser especially for the Kindle Fire, so you'd expect it to be easy to use and as fast as possible. Silk is certainly easy to use, but it's often not as fast as it might be.

 You can close all the tabs in the Silk browser at once by tapping and holding on a tab, and then tapping the Close All Tabs button in the dialog box that opens. If you want to close all tabs apart from the one you're viewing, tap the Close Other Tabs button in this dialog box instead.

If you find Silk running too slowly for your liking, follow these steps to get it running faster:

1. Open Silk if it's not currently running. For example, tap the Home button to display the Home screen, and then tap the Web button to open Silk.
2. Tap the Menu button to display the Menu panel.
3. Tap the Settings button to display the Settings screen.
4. Scroll all the way down to the bottom of the screen, so you can see the settings shown in Figure 4-9.
5. If the Enable Flash button shows On, tap the down-arrow button to display the Enable Flash dialog box (shown here). Then tap the Off button.

Enable Flash	
Always on	○
On demand	○
Off	●
Cancel	

6. Clear the Accelerate Page Loading check box.
7. If the Desktop Or Mobile View button says Automatic: Optimize For Each Website or Desktop: Optimize For Desktop View, tap the down-arrow button to display the Desktop Or Mobile View dialog box (shown here). Then tap the Mobile: Optimize For Mobile View option button.

Desktop or mobile view	
Automatic: Optimize for each website	○
Desktop: Optimize for desktop view	○
Mobile: Optimize for mobile view	●
Cancel	

8. Tap the Back button to return from the Settings screen to the Silk screen.

Now try some browsing and see if the webpages load more snappily.

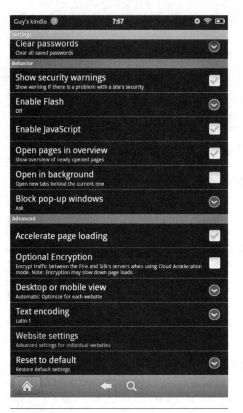

FIGURE 4-9 You can speed up the Silk browser by disabling Flash, turning off the Accelerate Page Loading feature, and choosing the Optimize For Mobile View setting.

DOUBLE GEEKERY

Understand What the Accelerate Page Loading Feature Does

Accelerate Page Loading is a feature that Amazon made a lot of noise about, but at this writing it seems only partially successful—so you may do better to turn it off.

When Accelerate Page Loading is turned on, the Silk browser requests content from Amazon Web Services rather than directly from the website that hosts the content. Amazon Web Services compresses the content so that it can get it to the Kindle Fire more quickly. Amazon Web Services caches a large amount of content from popular sites so that this content is ready to go when your Kindle Fire requests it, rather than Amazon Web Services having to go get the content and then serve it up to Silk.

Project 29: Squeeze the Battery Life to the Limit

Like all portable electronics, your Kindle Fire relies heavily on its battery. The battery is a lithium-ion model with an electrical capacity of 4400 milliamp-hours (mAh); it's built into the Kindle Fire, and it's not intended to be user-replaceable.

Amazon suggests the battery will give eight hours of playback, but if you use your Kindle Fire vigorously, you may find you get less.

See Exactly How Much Battery Life There Is

The battery icon in the status bar gives you a rough idea of how much power the battery contains. To get a more exact readout, follow these steps:

1. Tap the Home button to go to the Home screen.
2. Tap the Settings icon on the status bar to display the Settings icons.
3. Tap the More button to display the Settings screen.
4. Tap the Device button to display the Device screen.
5. Look at the Battery readout, as shown here.

Guy's kindle	3:26	⚙ 🛜 🔋
Device		

Application Storage (0.96 GB of 1.17 GB available)
Apps you've installed on your Kindle Fire

Internal Storage (5.14 GB of 5.37 GB available)
Content including Newsstand, Books, Music, Docs and Videos you've stored on your Kindle Fire

Battery 90% Remaining

System Version
Current Version: 6.2.2_user_3205220

DOUBLE GEEKERY

Understand What Charge Cycles Are and Why They're Important

Lithium-polymer batteries such as that used in the Kindle Fire are typically good for around 500 charge cycles before they start losing their capacity. But what *is* a charge cycle?

Each time you charge the battery up all the way and run it down all the way, that's one complete charge cycle. If you use just part of the battery, and then charge it up again, that's part of a charge cycle.

You can safely keep topping up the battery with partial charges. This is because with lithium-polymer batteries, you don't have to worry about the "memory effect," in which discharging the battery only partially before recharging it could reset the battery's Empty threshold to the point at which you stopped discharging it. This was a big problem with older types of rechargeable batteries—for example, resetting the Empty threshold to the 50-percent level would severely reduce the battery life.

Reduce Your Power Consumption

If you're not getting enough life out of the battery, you need to reduce the amount of power you're consuming. You can do this in several ways—some obvious, some not—explained in this section.

Reduce the Display Brightness

Your Kindle Fire's bright screen goes through a lot of power compared to the other components, so if you can bear to decrease the brightness, you can save power and extend battery life.

To reduce the display brightness, follow these steps:

1. Tap the Settings button on the status bar to display the settings bar.
2. Tap the Brightness button to display the Brightness slider (shown here).
3. Tap the Brightness slider button and drag it to the left.
4. When you finish setting the brightness, tap the Settings button to hide the settings bar again.

Turn Off Wi-Fi Unless You're Using It

Your Kindle Fire's next greediest feeder at the battery buffet is the Wi-Fi feature. So when you don't need Wi-Fi, turn it off rather than just letting it eat away at the battery life. For example, if you're reading a book or listening to music, you probably don't need Wi-Fi running.

To turn Wi-Fi off, follow these steps:

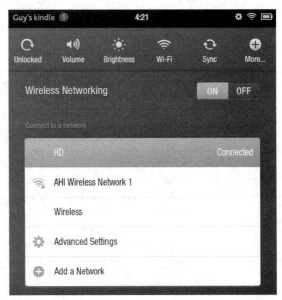

1. Tap the Settings button on the status bar to display the settings bar.
2. Tap the Wi-Fi button to display the Wi-Fi controls (shown here).
3. Tap the Wireless Networking switch and move it to the Off position.
4. Tap the Settings button again to hide the settings bar.

If you've set a restrictions password for Wi-Fi, when you turn Wi-Fi back on by tapping the Wireless Networking switch and moving it to the On position, your Kindle Fire prompts you to enter the password, as shown here. Type the password and then tap the OK button.

Set Your Kindle Fire to Go to Sleep After a Short Delay

For those times when you don't put the Kindle Fire to sleep manually, make sure it's putting itself to sleep after a short delay rather than staying awake. To set the sleep delay, follow these steps:

1. Tap the Home button to go to the Home screen.
2. Tap the Settings icon on the status bar to display the Settings icons.
3. Tap the More button to display the Settings screen.
4. Tap the Display button to display the Display screen (shown here).
5. Tap the Screen Timeout button to display the Screen Timeout screen (see Figure 4-10).
6. Tap the button for the timeout you want—for example, 30 Seconds, 1 Minute, or 5 Minutes.
7. Tap the Back button at the bottom of the screen to return to the Display screen.

DOUBLE GEEKERY

Should I Put the Kindle Fire to Sleep or Turn It Off Fully?

If you're desperate to save every milliamp of power you can, turn your Kindle Fire off fully. That way, it uses no power at all.

But under normal conditions, you're better off putting the Kindle Fire to sleep than turning it off fully. When asleep, the Kindle Fire uses only a minimal amount of power—and you can resume using it and get to work within seconds.

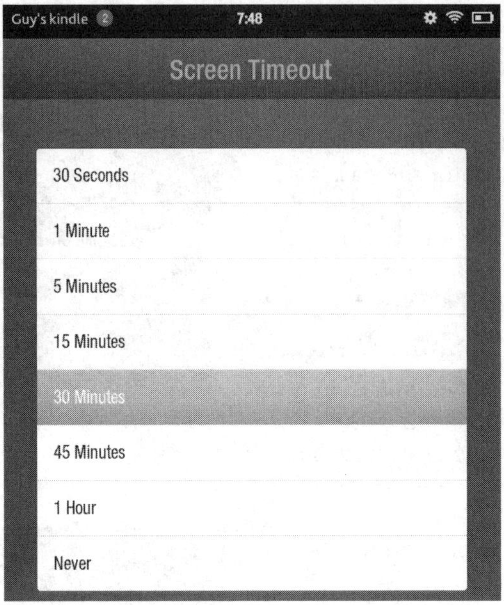

Guy's kindle ②   7:48   ⚙ 📶 🔋

Screen Timeout

30 Seconds

1 Minute

5 Minutes

15 Minutes

30 Minutes

45 Minutes

1 Hour

Never

FIGURE 4-10   On the Screen Timeout screen, set a short interval for the screen to time out and put the Kindle Fire to sleep.

Avoid Extremes of Heat and Cold

To get the best out of the Kindle Fire's battery, try not to let it get too hot or too cold.

Generally, if you keep your Kindle Fire with you, it'll be at least as happy as you are. I doubt you'll be tempted to use your Kindle Fire in the sauna or in the snow, but be careful when leaving your Kindle Fire in your car, where the temperature can soar in sunshine or plummet in winter.

If the Kindle Fire gets very hot or very cold, put it somewhere more temperate and let it return to a more normal temperature before trying to use it.

DOUBLE GEEKERY

Condition Your Kindle Fire's Battery

To enjoy the Kindle Fire to the full, you'll want to get the most life possible out of its battery. To do this, you need to condition the battery, which takes some time and patience up front.

To condition the battery, give it three full charges and discharges at the beginning of its life:

- **Charge the battery all the way up to 100 percent**   Plug the Kindle Fire into a USB power adapter and leave it to charge until it's full.
- **Run the battery all the way down until the Kindle Fire shuts itself down**   The easiest way to do this is to wind up the screen brightness all the way to high and set a marathon playlist of videos running. You can also simply use the Kindle Fire normally as long as you don't plug it into power.
- **Repeat the process twice**   Plug the Kindle Fire back into the USB power adapter, leave it until the battery is fully charged, and then run the battery down all the way again. Do that once more, and the battery should be in good shape.

This conditioning technique is simple enough to describe, but it takes real patience to do, because it takes two or three days—the Kindle Fire can usually manage five to eight hours of playback or other vigorous activity, and takes four to six hours to recharge fully, so each cycle takes the better part of a day. This tends to be grueling if you want to start using the Kindle Fire right the moment you get it. But if you can manage the three full charges and discharges, you'll get the best performance out of the battery.

After conditioning the battery, you can plug the Kindle Fire into a power source (a computer or the USB power adapter) whenever you need to. Lithium-ion batteries don't suffer the "memory effect" that plagued older battery technologies, in which charging the battery when it was only partly discharged could take a hefty chunk out of the battery's capacity to store power.

While we're on the subject, three more quick things on charging:

- Don't leave the Kindle Fire plugged into power for more than a day or two at a time, as doing this can shorten the battery life. Normally, this isn't a problem, as you'll likely be using the Kindle Fire out and about rather than chaining yourself to a socket.
- Unplug the USB power adapter from the electric socket when you're not using it. This helps both keep the power adapter happy and your electric bill down.
- At least once a month, run the Kindle Fire's battery all the way down, and then charge it all the way up. Lithium ion batteries like this treatment and give better performance as a result.

See Which App Is Chewing Through the Battery

Some apps are greedier for power than others, and if you find the power draining away unexpectedly fast, it's worth having a look to see what's consuming the power.

The Kindle Fire doesn't have a built-in way of checking power consumption the way some other Android tablets do—but you can add an app at no cost that does let you check power consumption. Follow these steps:

1. Tap the Home button to display the Home screen.
2. Tap the Apps button to display the Apps screen.

3. Tap the Store button to display the Store screen.
4. Tap in the Search In Appstore box at the top and type **badass**.
5. Tap the Badass Battery Monitor For Kindle Fire search result to display the list of matching items.
6. Tap the price button on the Badass Battery Monitor For Kindle Fire result, and then tap the Get App button. Your Kindle Fire downloads and installs Badass Battery Monitor.
7. Tap the Open button to open Badass Battery Monitor (see Figure 4-11).
8. To see the detail for a particular feature, tap the > button at the right end of its button. For example, if you tap the Wifi Active button, you'll see a screen such as that shown here.

Guy's kindle ④	11:19 ⚙ 📶 🔋

Wifi Detail - 1h 36m 18s (Since Last Unplugged)

🛡 **Wifi Active:** | **Wi-Fi Settings**

Usage Details

Wifi Active:	1h 12m 17s
Sent:	21.60 KB
Received:	788.51 KB
Total Bytes From Apps:	1.91 MB

Power Savings Hint

Turn off Wi-Fi when not in use using the Wi-Fi Settings button above.

9. From here, you can tap a command button (such as the Wi-Fi Settings button shown in the illustration) to go to the relevant controls or tap the Back button at the bottom of the screen to return to the main Badass Battery Monitor screen.
10. To see which particular apps have been using the most battery life, tap the > button at the right end of the App Usage button. Your Kindle Fire displays the screen shown in Figure 4-12, which lists the apps and system services by descending percentage of power used, so that the hungriest apps and services are at the top.

 After displaying the screen shown in Figure 4-12, you can tap the drop-down button at the right end of the silver bar near the top of the screen and then choose a different metric. For example, you can tap the View CPU Minutes Used option button to see which app or service has been using the processor most, or tap the View Network Data Used option button to see which app or service has been shifting the most data across the network connection.

Charge Your Kindle Fire While You're Out and About

If you prefer to use your Kindle Fire with the screen on full brightness and with Wi-Fi on, you may need to take a different approach to power management. Instead of rationing power and spoiling your tablet experience, you can refuel the Kindle Fire wherever possible.

Here are four suggestions for keeping your Kindle Fire adequately juiced while you're out and about:

- **Get an extra USB charger for work or school** Amazon sells a Kindle Fire Charger/AC Adapter for $20 or $25. Rather than carry your original charger everywhere, get an extra charger and keep it at work, at school, or wherever you need to charge the Kindle Fire regularly.

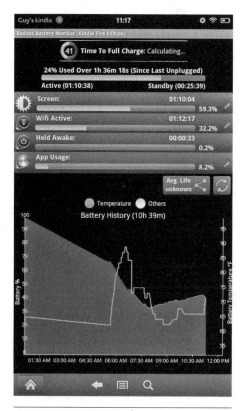

FIGURE 4-11   Use an app such as Badass Battery Monitor to see which apps and features are consuming most power on your Kindle Fire.

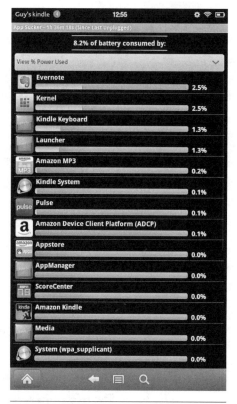

FIGURE 4-12   Display this screen to see which apps and system features are using the most battery power.

The Kindle Fire needs a current of at least 1800 milliamps (1.8 amps) to charge it. For best results, get a charger that gives 2100 milliamps (2.1 amps).

- **Get a car charger**   If you spend time in your car, get a car charger that'll charge the Kindle Fire from your car's accessory socket. If you're shopping on Amazon.com, the AmazonBasics 2-Port USB Car Charger With 2.1 Amp Output is a good value, but you'll also need to get a USB cable from 2.0A Male to Micro B to convey the power to the Kindle Fire.
- **Get a case with a built-in battery**   If you need serious amounts of battery power on the go, look for a case with a built-in battery. At this writing, the main contender is the Nyko Extended Battery Kindle Fire Power Case (around $80; www.nyko.com or various online stores), but by the time you read this, there should be plenty more.
- **Get an external battery**   For between $20 and $60, you can buy an external battery with one or more USB ports that'll deliver enough power to recharge your Kindle Fire once or twice. A typical external battery of this type is about half the size and weight of the Kindle Fire, so it'll fit easily enough in your pocket, purse, or tote.

Project 30: Prevent Your Children from Going Online with the Kindle Fire

If you have kids, you can be pretty sure they'll love the Kindle Fire and want to use it anytime you're not. This should be fine—but you'll need to make sure it stays that way.

Out of the box, the Kindle Fire is largely safe for kids to use—except for two problems:

- **Internet access**   The Kindle Fire's Silk browser doesn't have content restrictions, so anyone using the browser can delve into the full range of content that the Internet provides.
- **Your Amazon account**   Because the Kindle Fire doesn't have separate user accounts, anybody who borrows your Kindle Fire is working within your user account. If you've set up your user account with one-click purchasing from Amazon, anybody who picks up your Kindle Fire can spend every last remaining cent of your credit.

You can help protect your children's innocence and preserve your credit rating by preventing your children from going online with your Kindle Fire. That leaves them plenty to do with the Kindle Fire offline—they can read books, listen to the music stored on the Kindle Fire, watch videos you've put on it, and so on.

Protect Your Kindle Fire's Wi-Fi with a Restrictions Password

To prevent your kids from going online with your Kindle Fire, you can protect its Wi-Fi with a password, and then disable Wi-Fi. Unless your kids can figure out the password, they won't be able to connect the Kindle Fire to a network or access the Internet.

To protect your Kindle Fire in this way, follow these steps:

1. Tap the Settings button on the status bar to display the settings bar.
2. Tap the More button on the settings bar to display the Settings screen.
3. Tap the Restrictions button to display the Restrictions screen (shown here).

4. Tap the Enable Restrictions switch and move it to the On position. Your Kindle Fire displays the Create Password screen (shown here).

5. Type the password in both the Enter Password box and the Confirm Password box.

6. Tap the Finish button. Your Kindle Fire displays the remaining controls on the Restrictions screen (see Figure 4-13).

7. Tap the Password Protected Wi-Fi switch and move it to the On position. Your Kindle Fire displays the Restrictions Password screen (shown here).

8. Type your restrictions password, and then tap the OK button. Your Kindle Fire returns you to the Restrictions screen, where the Wireless Network button now shows the readout Wi-Fi Is Off.

The Kindle Fire is now safe for your kids to use.

FIGURE 4-13   After you move the Enable Restrictions switch to the On position and enter a restrictions password, your Kindle Fire displays the remaining controls on the Restrictions screen.

Turn Wi-Fi Back On

When you've got the Kindle Fire safely back in your paws and you're ready to turn Wi-Fi on again, go back to the Restrictions screen, and then tap the Wireless Network button. On the Wireless Network screen (shown here), tap the Wireless Network switch and move it to the On position.

Your Kindle Fire then displays the Restrictions Password screen. Type your restrictions password, and then tap the OK button.

Project 31: Turn Off One-Click Buy on Your Kindle Fire

Amazon has designed the Kindle Fire so that you can buy a book, CD, video, app, or other shiny digital bauble with a single click—well, a single tap—on the touch screen.

This is great for rampant consumers with inexhaustible petrodollar wealth at their disposal, but for the rest of us, One-Click Buy can be a bane rather than a boon. If you (like me) run the risk of digital lust getting the better of your willpower, you may want to turn off One-Click Buy to reduce the risk of ill-considered acquisitions.

To turn off One-Click Buy, follow these steps:

1. On your computer, open your browser and go to the Amazon.com website.
2. If your browser doesn't sign you in automatically, click the Sign In link, and then click the Sign In button. On the webpage that appears, type your e-mail address and password, and then click the Sign In button.
3. Click the Your Account drop-down button, and then click the Your Account link to display the Your Account screen.
4. In the Settings area, click the 1-Click Settings link to display the Manage Addresses And 1-Click Settings screen.
5. Click the Turn 1-Click Off button.

Project 32: Protect Your Kindle Fire Against Malware

Because your Kindle Fire can access the Internet, you may need to protect it against malevolent software, or *malware. May?* Yes—at this writing, there are few viruses and malware programs that target Android systems, and if you keep your Kindle Fire within the Amazon ecosystem of approved apps, you may be safe enough. And you can avoid the performance hit and decreased battery life that antivirus software tends to bring while monitoring your Kindle Fire in real time.

But you're reading this book, so we know you're prepared to take your Kindle Fire for a walk—or many—on the wild side. If you side-load apps (as explained in Project 34), you run the risk of installing an app that appears to be friendly but actually targets your data. Such apps may act as Trojan horses, containing malevolent code within an app that actually does something useful or amusing.

And, of course, things move quickly in our wired world. A few years ago, there were pretty much no viruses that targeted OS X. Now there are some—not as many as there are for Windows (which remains the main target of virus writers)—but quite enough. Soon, virus writers may target Android as well.

So you may well decide that you should install an antivirus app. This project introduces you to one of the leading contenders, Dr. Web Anti-Virus Light. This is a free app you can download from the Appstore, but it seems to provide pretty good protection.

 You'll find other antivirus apps in the Appstore, and you may want to evaluate them all. If you do, make sure you run only one antivirus app at a time. Otherwise, they may disagree with each other (in real time).

Download and Install Dr. Web Anti-Virus Light

To get started, download and install Dr. Web Anti-Virus Light. Follow these steps:

1. On the Home screen, tap the Apps button to display the Apps screen.
2. Tap the Store button to get to the Appstore.
3. Search for **dr. web**, and then tap the search result.
4. Tap the Free button, and then tap the Get App button to start the installation running.

Meet the Dr. Web User Interface

When the installation finishes, tap the Open button to open Dr. Web. You'll then see the Security Center screen (see Figure 4-14), which gives you access to the main controls.

If the button to the right of SpIDer Guard shows Off, tap it to turn the monitor on.

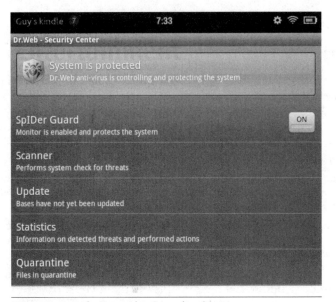

FIGURE 4-14 From the Security Center screen, you can run a scan, update the app's databases of threats, view statistics on threats detected and actions taken, or check the files Dr. Web has put in quarantine.

Run a Scan

To run a scan, follow these steps from the Security Center screen:

1. Tap the Scanner button to display the Scanner screen (shown here).

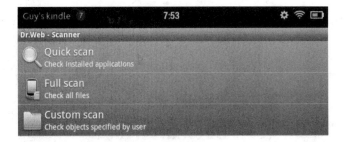

2. Tap the scan type you want:
   - **Quick Scan**   Tap this button to check the apps on your Kindle Fire.
   - **Full Scan**   Tap this button to check everything. This is what you'll probably want to do at first.
   - **Custom Scan**   Tap this button when you want to scan only particular folders.

3. Wait while the scan runs. Dr. Web keeps you up to date, as shown here.

4. When the scan completes, follow the instructions to deal with any threats identified.

Update the Dr. Web Database of Threats

To make sure your Kindle Fire is protected against new threats, update Dr. Web's database of threats. To do so, tap the Update button on the Security Center screen. Dr. Web downloads the latest files, and then applies them.

View Statistics on Threats Detected and Actions Performed

To see which threats Dr. Web has detected and which actions it has performed (such as putting rabid files into quarantine), tap the Statistics button on the Security Center screen. Dr. Web displays the Statistics tab of the Statistics/Quarantine screen (see Figure 4-15).

From here, you can either tap the Quarantine tab to display its contents (discussed in the next section) or tap the Back button to go back to the Security Center screen.

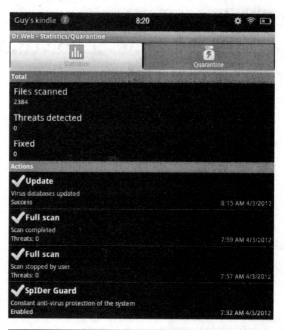

FIGURE 4-15   Look at the Statistics tab of the Statistics/Quarantine screen to see which threats Dr. Web has detected and fixed.

View the Files Dr. Web Has Quarantined

To see which files Dr. Web has quarantined, tap the Quarantine button on the Security Center screen. If you're currently viewing the Statistics tab of the Statistics/ Quarantine screen, tap the Quarantine tab. Either way, Dr. Web displays the Quarantine tab, shown here without any files quarantined.

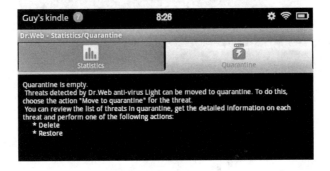

If Dr. Web has quarantined files, look through them and decide how to handle them. You can either tap the Delete button to delete a file or the Restore button to restore it to where Dr. Web collared it.

Project 33: Get Apps from Other Sources than Amazon's Appstore

As you know, tapping the Store button on the Kindle Fire's Apps screen takes you to Amazon's Appstore, which provides a selection of apps. These apps are vetted both for their system demands and for their content, so you can be sure not only that they'll run on your Kindle Fire but also that you won't get any unpleasant surprises.

But because your Kindle Fire runs the widely used Android operating system, you can also install many other apps that you can find on the Internet.

In this project, I'll show you how to get apps from other sources. In the next project, you'll learn how to install them on your Kindle Fire using a technique called side-loading.

Android apps come in application package files that use the .apk file extension and are often referred to as APKs. So your first move is to get the package files for the apps you want to install on your Kindle Fire. There are three main ways of getting them:

- Download the package file from Android Market to an Android device, and then extract the file.
- Download the package file from a website.
- Download the package file from other online sources.

The following sections walk you through these three approaches.

What's JAR Got to Do with It?

The APK file format that Android app package files use is based on the Java Archive file format, JAR for short. JAR is a special kind of Zip file used for distributing Java application files, so APK is also a kind of Zip file.

This relationship between APK and JAR is why "jar" appears in the names of various sites that distribute APK files, such as GetJar (www.getjar.com).

Get Package Files from Android Market Using an Android Device

If you have a regular Android device, you can get app package files from Android Market, extract the files to your computer, and then load them on your Kindle Fire.

Side-loading the apps onto the Kindle Fire as described in the next project doesn't allow paid apps to authorize properly, even apps you've bought using your own account on Android Market. So you're normally better off using free apps for side-loading on the Kindle Fire—but see the nearby sidebar for a warning about apps that are free because they're supported by ads.

Android Market is Google's official source of Android apps, so this is the best approach if you have a regular Android device—one that's not limited in the ways that the Kindle Fire is. For example, if you have a Samsung Galaxy Tab, a Motorola Xoom, or one of the many smart phones that runs Android, you can get apps from Android Market.

It *is* possible to install directly apps from Google's Android Market on your Kindle Fire—but before you can do so, you must "root" your Kindle Fire. That topic is outside the scope of this book, but the curious can find instructions and utilities on the Web.

Understand the Threat from Ad-Supported Free Apps

Free apps are great—but you should be careful of apps that are free because they're supported by ads.

Mostly, the ads are harmless, although some can be irritating. But because of the way the Android operating system is designed, the ad libraries—the software that operate the ads—get the same level of permissions as the apps themselves. So when you're installing an app, and you get the chance to review the permissions an app needs, keep in mind that the ad libraries will have these permissions too.

For example, the app shown in the illustration (which will remain nameless because it's blameless) requires the permission "Read the user's contact's data." That means the app requires access to your address book, and the ad libraries will have access to this information too—and can share it online. I doubt you want this.

Three quick things before we go on. First, lots of apps want address book access for social networking purposes—they want to know whom you know. Second, you can see the "Write (but not read) the user's contacts data" permission right below the "Read the user's contacts data" permission. These two permissions seem contradictory, but from a system point of view, it's vital to be able to grant the write permission separately. Third, the "Broadcast sticky intents" permission sounds lubricious but merely means that after the app broadcasts information, the system stores it so that other apps can access it without having to wait for the next broadcast. This is normal and harmless.

The more information an advertiser can learn about you, the better the advertiser's chance of making money from you. So information such as your address or location is especially valuable.

So by using ad-supported apps, you may be risking your privacy—and possibly your security—for a handful of dollars. If you find the app so useful you expect to use it extensively and it costs only a few dollars, you will be better off spending that money than compromising your privacy.

Get an App from Android Market

To get an app from Android Market using a standard Android device, follow these general steps. The steps vary somewhat depending on the device, which version of Android it's running, and whether that version has been customized—but you should have no problem following them.

1. From the Home screen of your Android device, tap the Apps button to display the Apps screen.
2. Tap the Market icon to open the Market app. The Market app displays the Apps screen (see Figure 4-16), which contains various lists such as the Staff Picks list, the Top Paid list, and the Categories list. You can also search for an app by tapping in the Search box, typing your search terms, and then tapping the ENTER key.
3. Navigate to the app you want, either by browsing or by searching, tap its Install button, and then follow through the process for installing the app. Figure 4-17 shows a search that has found ES File Explorer.

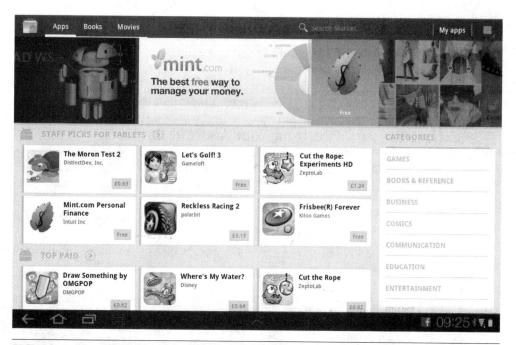

FIGURE 4-16   On the Apps screen in Android Market, you can browse through the apps by categories or search for the apps you want.

FIGURE 4-17   If you know the name of the app you want, searching is usually the quickest way to locate it.

Use this technique to search for the free ES File Explorer app and install it on your Android device. You'll need ES File Explorer in the next section to extract the APK file for the app you want to put on your Kindle Fire.

If you have another file manager, such as Astro File Manager, on your Android device, you can use that file manager instead.

Extract the APK File from Your Android Device

After installing the app on your Android device, you can extract the APK file from the device so that you can install the app on your Kindle Fire.

To extract the APK file, you use ES File Explorer. Follow these steps:

1. From the Home screen of your Android device, tap the Apps button to display the Apps screen.
2. Tap ES File Explorer to launch the app. You'll see the folders on your Android device, as shown here.

3. Tap the AppMgr button on the toolbar to display the AppMgr screen. The next illustration shows the top part of the AppMgr screen.

4. Tap and hold the app whose APK file you want to get. ES File Explorer displays the Operations dialog box (shown next).

▣ Operations
Select All
Uninstall
Backup
Shortcut
Detail
Restore
Help Center

5. Tap the Backup button. ES File Explorer backs up the file and briefly displays a pop-up message saying where it has put the backup, as shown here.

Dolphin Browser HD_7.3.0.apk was backed up successfully (/sdcard/backups/apps/)

08:10

6. Tap the Back button on the toolbar to display the list of folders again.
7. Navigate to the folder that contains the backup.
8. Transfer the backup to your computer in one of these ways:
   - **Direct connection**   Connect your Android device to your PC or Mac and simply copy the file across using Windows Explorer or the Finder.
   - **E-mail**   Attach the file to a message and send it to yourself.
   - **SD card**   If your Android device has an external SD card, copy the backup file to the SD card. Then put the SD card in your computer and copy it across.
   - **File-sharing service**   If you use Dropbox or another file-sharing service, upload the backup file to your Dropbox account from your Android device. The file then appears in the Dropbox folder on your computer.

Download Package Files from a Website

If you don't have an Android device that you can use to download the APK file for the app you need, you may be able to download the package file from a website.

Here are three websites that provide a wide range of APK files:

- **GetJar**   GetJar (www.getjar.com) has a wide range of different files sorted into categories such as Games, Social & Messaging, Productivity, Education, Health, and Religion.
- **AndroidShock**   AndroidShock (www.androidshock.com) is a games site that has both free games and pay games.
- **Freeware Lovers**   Freeware Lovers (www.freewarelovers.com) is a freeware site that has Android apps in categories including Communications, Finance, Multimedia, Productivity, and Travel.

Download Package Files from Other Sources

You can also download package files from other sources online. For example, if you use a peer-to-peer (P2P) networking client, you can find many Android package files. Many of these are distributed illegally, so you may prefer not to download them.

Project 34: Side-Load Apps on Your Kindle Fire

After getting the APK file for the app you want to install on your Kindle Fire, you can side-load the app using the procedure explained in this project. *Side-loading* means installing apps by transferring them to the Kindle Fire via USB.

Install ES File Explorer

Your first step toward side-loading apps onto your Kindle Fire is to install ES File Explorer on your Kindle Fire. This is the same app you installed on your other Android device in the previous project to extract the APK file from the device to a file on your PC. ES File Explorer is a file manager that lets you manipulate files on your Kindle Fire in much the same way you'd manipulate them on your PC or Mac.

ES File Explorer is free and is available from Amazon's Appstore. To install ES File Explorer, follow these steps:

1. Tap the Home button to display the Home screen.
2. Tap the Apps button to display the Apps screen.
3. Tap the Store button to display the Store screen.
4. Tap in the Search In Appstore box to place the insertion point there.
5. Type **es file** and wait while the Kindle Fire returns matching search results.
6. Tap the ES File Explorer For Kindle Fire item to display the list of matching apps.
7. Tap the FREE button on the ES File Explorer item to display the app's screen.
8. Tap the FREE button.
9. Tap the Get App button. Your Kindle Fire then installs ES File Explorer.

Allow the Installation of Apps from Other Sources

Next, you need to make your Kindle Fire allow the installation of apps from sources other than Amazon.com. Follow these steps:

1. Tap the Settings button on the status bar to display the settings bar.
2. Tap the More button on the settings bar to display the Settings screen.
3. Tap the Device button to display the Device screen (see Figure 4-18).
4. Tap the Allow Installation Of Applications From Unknown Sources switch and move it to the On position.
5. Tap the Back button twice to go back the screen you started from.

Copy the Package Files to Your Kindle Fire

Now copy the package files to your Kindle Fire. Follow these steps:

1. Connect your Kindle Fire to your computer using the USB cable.
2. Open a Windows Explorer window or a Finder window to the folder you put the APK file in.
3. Right-click (or CTRL-click on the Mac) the APK file, and then click Copy on the context menu.
4. Navigate to the Kindle Fire, and then paste the APK file into a convenient folder. You don't have to use a special folder, but you may find your Download folder a convenient place.
5. Eject the Kindle Fire from your computer and then disconnect the USB cable.

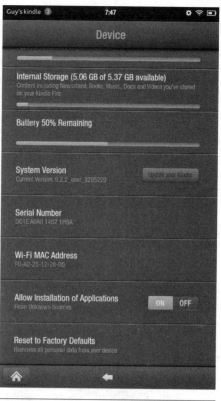

FIGURE 4-18   On the Device screen, tap the Allow Installation Of Applications From Unknown Sources switch and move it to the On position.

Install the Apps on the Kindle Fire

After you copy the package files for the apps from your computer to your Kindle Fire, you can install the apps using ES File Explorer. Follow these steps:

1. Tap the Home button to display the Home screen.
2. Tap the Apps button to display the Apps screen.

3. Tap the ES File Explorer button to launch ES File Explorer.
4. Navigate to the folder you put the package files in.
5. Tap the package file for the app you want to install. Your Kindle Fire displays a screen asking if you want to install the app, as shown in Figure 4-19.
6. Click the Install button to proceed with the installation. The installer then runs. When it finishes, it displays the Application Installed message.
7. Tap the Open button to open the app. The app then runs as normal. Figure 4-20 shows the Dolphin browser up and running.

Subsequently, you can run the app from the Apps screen, just like any app you've installed from Amazon's Appstore.

FIGURE 4-19 On the installation screen for an app you're side-loading, check the permissions the app needs, and then click the Install button.

FIGURE 4-20 After side-loading an app, you can run it from the Apps screen and use it as normal.

Project 35: Make the Most of Your Kindle Fire's Wi-Fi Capabilities

As you know, your Kindle Fire has two means of communicating with the outside world: USB and Wi-Fi. USB is great for transferring data to or from your PC or Mac, but Wi-Fi is often more convenient—especially if you mislay your USB cable.

In the section, I'll show you how to make the most of your Kindle Fire's Wi-Fi capabilities. We'll look at how to transfer files wirelessly, examine wireless networks, and add extra storage to your Kindle Fire.

Transfer Files Wirelessly Using Wi-Fi File Explorer

If you want to be able to access your Kindle Fire from your computer via Wi-Fi, download and install WiFi File Explorer from the Appstore.

WiFi File Explorer comes in two versions:

- **WiFi File Explorer**   This version is free, contains ads, and lacks some of the most useful features, such as creating folders on your Kindle Fire.
- **WiFi File Explorer Pro**   This version costs $0.99 at this writing, has no ads, and has full features.

Launch WiFi File Explorer on Your Kindle Fire and Connect with Your Computer

When you run WiFi File Explorer, the app displays the WiFi File Explorer dialog box shown here, giving the IP address where you can access the Kindle Fire.

DOUBLE GEEKERY

What's the ":8000" Bit at the End of the Address?

The **:8000** part after the IP address (in the example, **10.0.0.43**) is the port that WiFi File Explorer is using on your Kindle Fire. A *port* is a different channel (as it were) on the IP address. Different programs use different ports both for security and to avoid confusion.

When you access a regular URL, your browser automatically uses the default HTTP port 80, so you don't need to type it. But when you're accessing your Kindle Fire via WiFi File Explorer, you must type the port number. If you don't, your browser won't find WiFi File Explorer on the Kindle Fire.

Tap the Done button to close the WiFi File Explorer dialog box. You then see the WiFi File Explorer status screen (shown here), which shows the Wi-Fi status and the URL for reference.

Now open a browser window on your computer and type that address into the address box—for example, **http://10.0.0.43:8000**. You then see a WiFi File Explorer that shows the contents of the top level of your Kindle Fire's file system (see Figure 4-21). The top level is the root folder, which is represented by a / character.

Meet the WiFi File Explorer User Interface

WiFi File Explorer has a pretty straightforward user interface, with most of the buttons labeled. The next illustration shows the important buttons that aren't labeled:

- **Show/Hide Tree View Pane** Click this button to display or hide the Tree View pane on the left of the window.

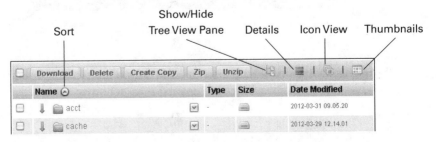

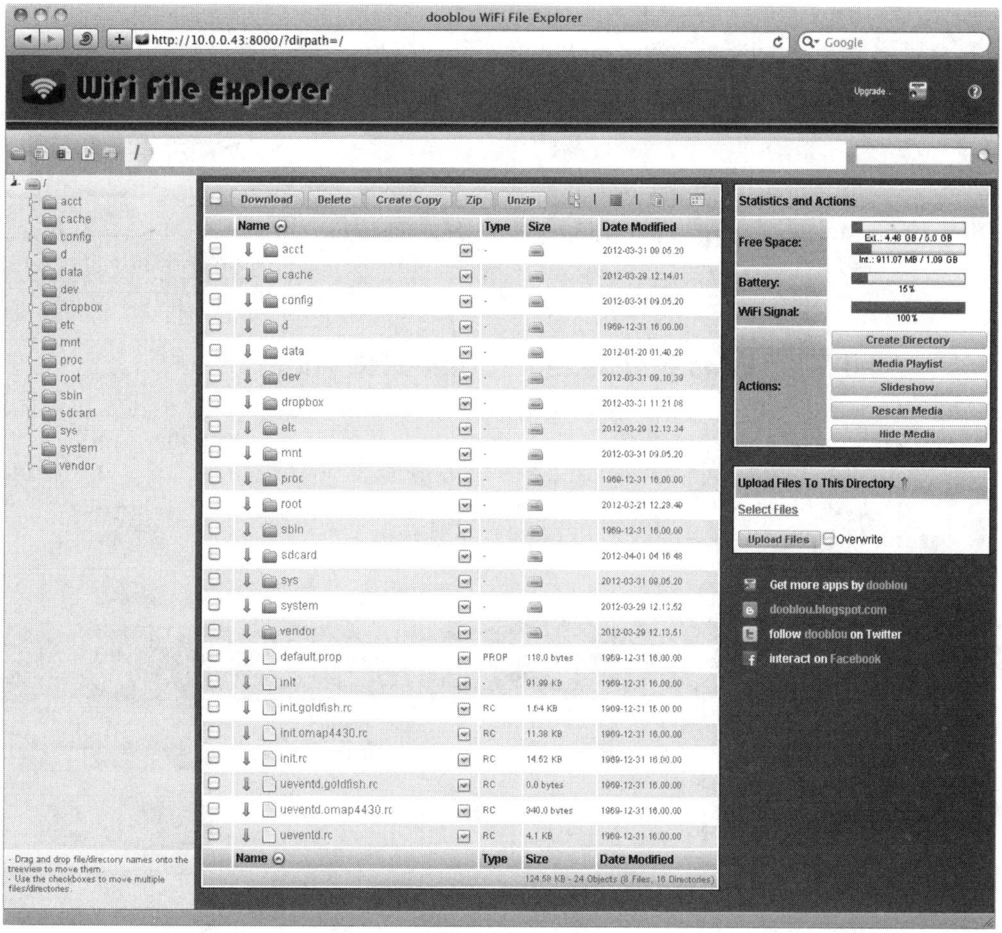

FIGURE 4-21   When you connect to your Kindle Fire via WiFi File Explorer, you first see the top level of the tablet's file system.

- **Details**   Click this button to display the files and folders listing in Details view, which you see in Figure 4-21.
- **Icon View**   Click this button to display the files and folders listing in Icon view, which displays an icon along with the detail of each item, as shown in the next illustration.

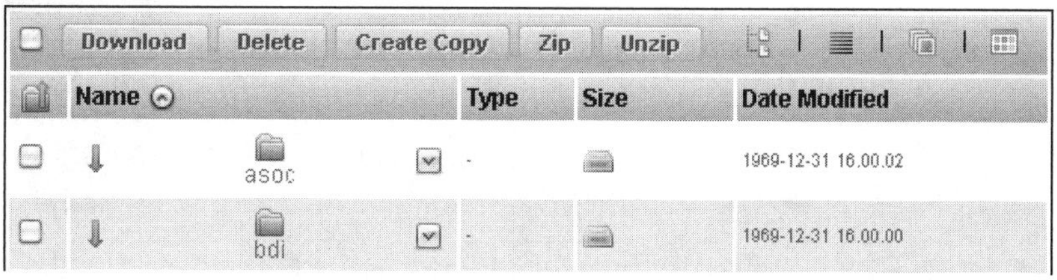

- **Thumbnails**   Click this button to display the files and folders listing as thumbnails, as shown in the next illustration.

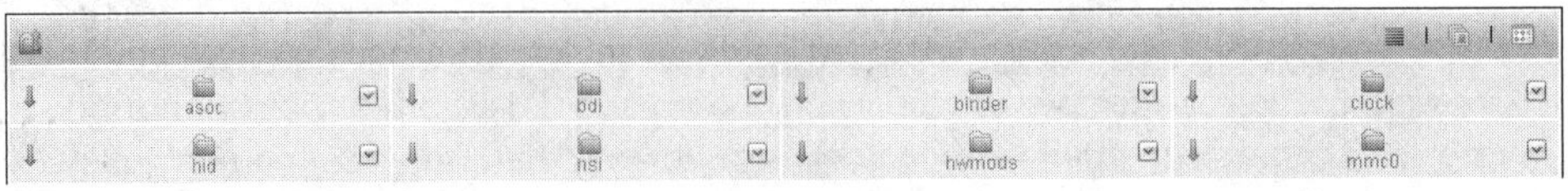

- **Sort**   Click the column heading by which you want to sort the list of files and folders—Name, Type, Size, or Date Modified. The Sort button indicates which column you're currently using for sorting. The default column is Name.

Find Your Files and Take Actions with Them

From the root folder, you can display the contents of any of the folders by clicking it either in the tree in the Tree View pane or in the list in the middle pane.

Where you'll probably want to spend most of your time is in the sdcard folder, which holds your files. Click the sdcard folder either in the Tree View pane or in the middle pane to display its contents (see Figure 4-22).

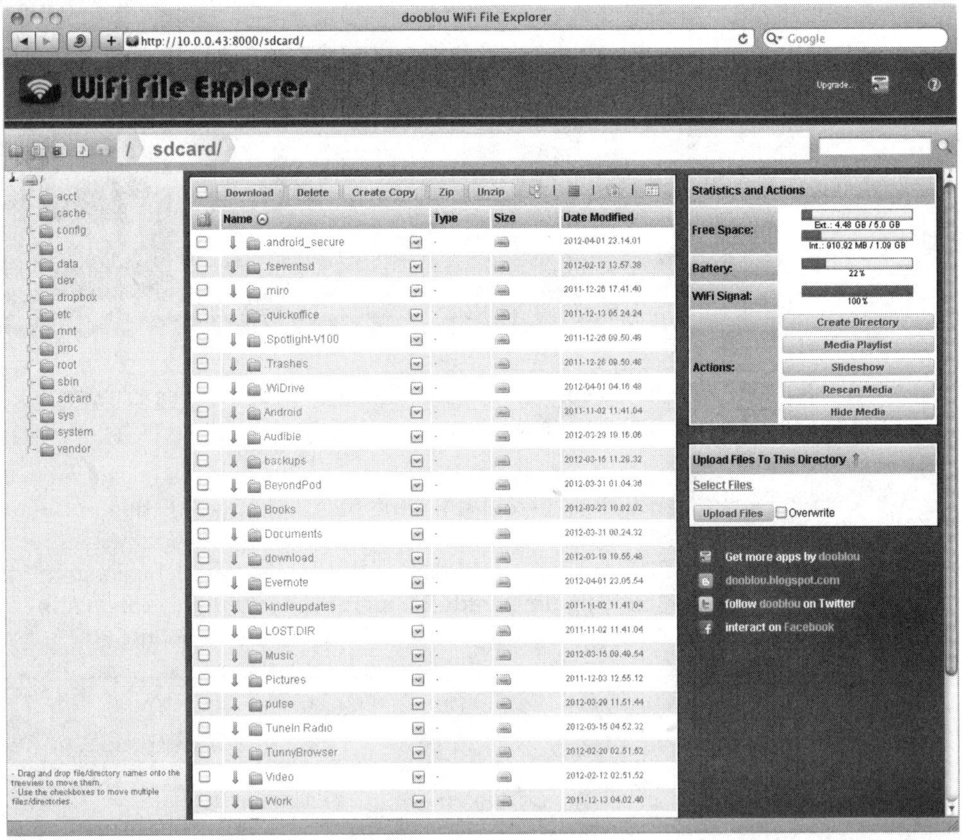

FIGURE 4-22   The sdcard folder contains most of the folders and files you'll want to work with—your Books folder, your Documents folder, your Music folder, and so on.

Now that you're in the right place, you can copy and move files and folders as needed:

- **Move a file or folder**   With the Tree View pane displayed, expand the tree so that you can see the destination folder. Then click the file or folder's name in the middle pane and drag it to the destination folder.
- **Copy a file or folder to your computer**   Click the down-arrow button to the right of the file or folder's name, and then click Download on the pop-up menu (shown here). Your browser downloads the file or folder to the folder it normally uses for downloads.

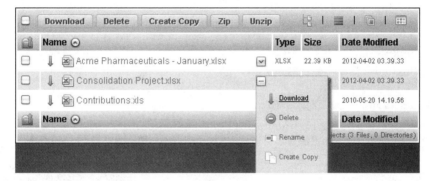

- **Copy files or folders to your Kindle Fire**   Follow these steps:
  1. Select the folder you want to put the files or folders in.
  2. On the right side of the WiFi File Explorer page, click the Select Files link in the Upload Files To This Directory box. Your browser displays the Select File(s) To Upload dialog box.
  3. Select the files or folders.
  4. Click the Open button. The WiFi File Explorer page lists the files in the Upload Files To This Directory box, as shown here.
  5. Select the Overwrite check box if you want the files to overwrite any existing files that have the same name.
  6. Click the Upload Files button. WiFi File Explorer uploads the files.
- **Delete a file or folder**
  Click the down-arrow button to the right of the file or folder's name, and then click Delete on the pop-up menu. In the confirmation dialog box that appears (shown here), click the OK button.

- **Work with multiple files or folders**   Select the check box for each file or folder you want to affect, and then click the appropriate button on the toolbar. For example, click the Download button to download the files or folders to your computer.
- **Create a folder**   Navigate to the folder in which you want to create the new folder, then click the Create Directory button in the Statistics And Actions box. In the dialog box that opens (shown here), type the folder name, and then click the OK button.

> http://10.0.0.43:8000
>
> Please enter the new directory name:
>
> Presentatations
>
> Cancel   OK

- **Rename a file or folder**   Click the down-arrow button to the right of the file or folder's name, and then click Rename on the pop-up menu. Type the new name in the dialog box that opens, and then click the OK button.

Close WiFi File Explorer

When you finish using WiFi File Explorer, tap the Menu button to open the Menu panel, and then tap the Exit button to close the app.

 If you want to leave WiFi File Explorer running in the background while you use other apps on your Kindle Fire, tap the Home button to display the Home screen instead of choosing Menu | Exit to close WiFi File Explorer.

Find Wi-Fi Networks Using Wi-Fi Analyzer

Your Kindle Fire's Wireless Network screen (see Figure 4-23) makes it easy to connect to Wi-Fi networks. But sometimes you'll want to know more about the wireless networks available before you try to join them.

To find out more about wireless networks, download and install the Wi-Fi Analyzer app from the Appstore. This app is free (supported by ads), and it's a great addition to your Kindle Fire.

When you run Wi-Fi Analyzer, it checks out all the Wi-Fi networks within reach and displays several helpful readouts. The top part of Figure 4-24 shows the relative signal strength of three networks, while the bottom part shows the speed for a wireless network.

FIGURE 4-23   From the Wireless Network screen, you can connect to a nearby network by tapping its name and providing the password or other security information needed.

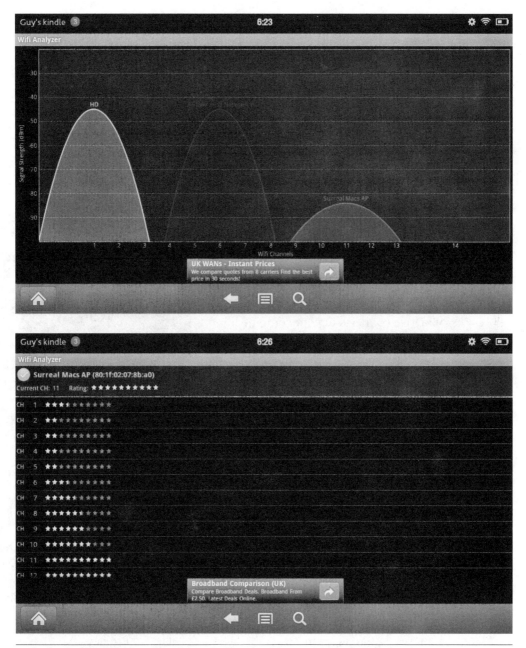

FIGURE 4-24   Wi-Fi Analyzer can help you choose which Wi-Fi networks to use. Readouts include the relative signal strength of available networks (top) and the channel distribution for a particular network.

Add Further Storage Using a Wi-Fi Drive

As you've seen earlier in this book, your Kindle Fire gives you around 5GB of space for storing the files you want to take with you—that's its 8GB nominal capacity minus the space needed for its operating system and virtual memory. That's enough for a decent number of songs plus a couple of movies—enough to keep you entertained as long as you can also access the other songs and movies you've stored on Cloud Drive.

But when you're traveling and can't access Cloud Drive, you can give your Kindle Fire extra storage space by using a Wi-Fi drive. These are portable drives—either hard drives or Flash drives—that include Wi-Fi capability for streaming files to devices such as your Kindle Fire.

Here are two of the leading Wi-Fi drives at this writing:

- **Kingston Wi-Drive** The Wi-Drive is a wireless access point with built-in storage—up to 64GB, depending on how much you're prepared to pay. The Wi-Drive is about the size and shape of an iPhone, so it's pretty portable. The Wi-Drive connects to your computer via a mini-USB port along which you load data and charge the drive. To use the Wi-Drive with your Kindle Fire, you install a companion app—also called Wi-Drive—and run it when you need a connection. Figure 4-25 shows Wi-Drive.
- **Maxell AirStash** The AirStash is a USB wireless access point that includes an SD card slot you can load with whichever size SD card you prefer (or whichever size you can afford). The AirStash is about the size of a large thumb, so it's easy to carry around. It has a built-in battery that recharges from a USB port and gives about six hours of life. At this writing, Maxell provides only an iOS app (called AirStash+) for accessing the AirStash, but you can connect to the AirStash from your Kindle Fire just as you connect to any other Wi-Fi network.

FIGURE 4-25 The Wi-Drive app enables your Kindle Fire to access a Kingston Wi-Drive Wi-Fi drive and stream files from it.

Project 36: Enter Text More Quickly in Documents and E-mail Messages

To make the most of your Kindle Fire, you'll want to enter text as quickly as possible in the documents and e-mail messages you create. To enter text quickly, you need to know and exploit the shortcuts and hidden features built into the Kindle Fire's keyboard.

> At this writing, you can't connect a hardware keyboard to the Kindle Fire without serious hardware hacking. The current Kindle Fire model has no Bluetooth capability, so you can't connect one of the many Bluetooth keyboards available. And although the Kindle Fire has a micro-USB port, it's not the kind of USB port to which you can connect a keyboard. So you're stuck with making the most of the on-screen keyboard.

To enter text as quickly as possible, use these features:

- **Caps Lock**   When you need to type in several letters or words all in capitals, tap the Shift button twice in immediate succession to turn Caps Lock on. The Shift button turns orange to indicate that Caps Lock is on. Type whatever you need in caps, and then tap the Shift button once to turn off Caps Lock.

- **Enter numbers using the top row of letters**   Tap and hold the top row of letters to get the numbers shown in miniature at the upper-right corner of each key. For example, tap and hold Q to display a panel containing a 1 button, as shown here, and then tap that button. Tap and hold W for 2, E for 3 (and accented e characters), and so on through P for 0 (zero).

- **Enter a period quickly**   Tap the spacebar twice in quick succession to enter a period and a space.

- **Enter punctuation quickly (part 1)**   To enter punctuation quickly, tap the 123!? button but don't lift your finger again. Instead, slide it to the punctuation key you want, and then lift your finger. The Kindle Fire enters the punctuation and displays the letters keyboard again without you needing to tap the ABC button.

- **Enter punctuation quickly (part 2)**   Another way to enter punctuation quickly is to use the punctuation panel. Tap and hold the period key to display the punctuation panel, slide your finger to the character you want, and then lift your finger.

- **Enter accented letters and variant letters quickly using the pop-up panels** To enter an accented letter (such as à) or a variant letter (such as the Spanish enye, ñ), tap and hold the base letter to display the pop-up panel of accented and variant letters, as shown here. Then tap the character you want to insert.

- **Make the most of the spelling suggestions** The spelling suggestions bar (shown here) seems clunky at first, but if you use it persistently, you can save plenty of typing.

- **Use Auto-Capitalization and Quick Fixes** The Kindle Fire can automatically capitalize the first letter of a sentence or paragraph for you, which is usually helpful. It can also fix a list of common typos, such as typing *hte* instead of *the*. To make sure you're using both these features, follow these steps:

  1. Tap the Settings button on the status bar to display the settings bar.
  2. Tap the More button to display the Settings screen.
  3. Tap the Kindle Keyboard button to display the Kindle Keyboard screen (see Figure 4-26).

FIGURE 4-26 On the Kindle Keyboard screen in Settings, make sure the Auto-Capitalization and Quick Fixes switches are set to the On position. You can also have the Kindle Fire play a sound to confirm each keypress.

DOUBLE GEEKERY

Type Faster Using the Swype Keyboard

If you need to type extensively on your Kindle Fire, try installing the Swype keyboard, which you can get by signing up at the Swype website (www.swype.com). This keyboard lets you enter letters by swiping your finger across the keyboard, changing direction at each letter you want, as shown here. Swype figures out matching words and presents a list.

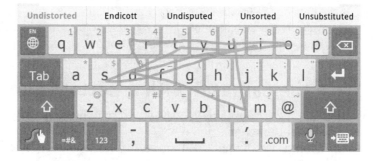

Using Swype feels awkward at first, but with practice, many people find it faster and more accurate than using the regular keyboard.

 From the Kindle Keyboard screen, you can also control whether the Kindle Fire plays a sound each time you tap a key on the keyboard. Some people find this helpful; others find it distracting. Move the Sound On Keypress switch to the On position or the Off position, as needed.

4. Make sure the Auto-Capitalization switch is in the On position.
5. Make sure the Quick Fixes switch is in the On position.
6. Tap the Back button to return to the Settings screen.

Project 37: Customize the Carousel—Or Replace It with a Better Launcher

The Carousel on the Home screen is an interesting way of opening recently used files and apps, but it doesn't suit everyone. In particular, you may not want the Carousel to show the items you've used most recently—especially if you share your Kindle Fire with other people.

In this project, we'll look first at how to customize the Carousel by removing items you don't want to appear. Then I'll show you how to replace the Carousel with a better launcher.

Remove Unwanted Items from the Carousel

When Amazon first released the Kindle Fire, there was no way to remove items from the Carousel. The resulting howls of protest quickly brought a system update that included an easy way of customizing the Carousel.

1. Tap and hold the item you want to affect until the pop-up menu appears (see Figure 4-27).
2. Tap the Remove From Carousel button to remove the item.

 If the item on the Carousel is seriously embarrassing (or useless), you can tap the Remove From Device button on the pop-up menu to remove it. If the item is an app, your Kindle Fire displays the Uninstall Application? screen to confirm the deletion.

Replace the Carousel with a Better Launcher

If you don't like the Carousel, or you simply want to know your choices, you can replace it with Go Launcher.

Go Launcher is free, but at this writing, it's not available in the Appstore. So you'll need to get Go Launcher using one of the methods discussed in Project 33: extracting the package file from an Android device that can access Android Market or downloading the package file from a website or other online resource. Then use the technique explained in Project 34 to side-load Go Launcher on your Kindle Fire.

When you run Go Launcher, you'll see the Home screen transformed. Figure 4-28 shows an example of how Go Launcher can look—but you can customize it widely by applying wallpaper, adding widgets, and choosing settings.

There's one more step to switch to Go Launcher: When you tap the Home button, you'll see the Complete Action Using dialog box, shown here. Tap the Use By Default For This Action check box to select it, and then tap the Go Launcher button.

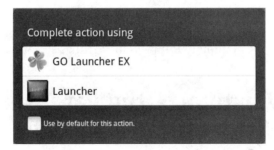

FIGURE 4-27 You can quickly remove unwanted items from the Carousel by using the Remove From Carousel button on the pop-up menu.

FIGURE 4-28 You can replace the Carousel on your Kindle Fire with Go Launcher to give a highly customizable Home screen.

Project 38: Recover Space from Your Kindle Fire's Storage

The Kindle Fire has only a modest amount of storage, so you'll need to actively manage your files to make sure you have enough space to load the apps, documents, and media files you need to take with you.

In this project, we'll look at three actions you can take to recover space:

- Deleting videos, pictures, songs, e-books, and documents you don't need
- Clearing out your Download folder
- Deleting package files for apps you've side-loaded
- Uninstalling apps you no longer need

Delete Unnecessary Videos, Pictures, Songs, E-books, and Documents

Start by browsing the following folders and seeing which files you no longer need:

- **Video** Video files take up a large amount of space, so start here. Delete any files you don't need.

 You can delete files from your Kindle Fire either by working on the tablet itself or by connecting it to your computer and using Windows Explorer or the Finder. If you're working on your Kindle Fire, you'll need a file manager such as ES File Explorer.

- **Pictures** Picture files can range in size from tiny to pretty big, so go through your pictures and see which you can dispense with. If you've loaded photos on your Kindle Fire, decide which to keep and which to lose.
- **Music** Music files, too, can be large, but normally it's the quantity of files that eats up your Kindle Fire's storage. Choose which songs you will keep with you and which you are happy to play from Cloud Drive or reload from your computer as needed.
- **Books** Many e-book files are so small that it's barely worth culling your library unless you've stuffed hundreds of e-books into it. But some books weigh in at 10 to 20MB, which is definitely enough space to recover.
- **Documents** If you've loaded many documents, or large documents (for example, presentations), on your Kindle Fire, see which you can get rid of for now.

Clear Out Your Download Folder

Open your Download folder in your file-management app (for example, ES File Explorer on your Kindle Fire, or Windows Explorer or Finder on your computer) and go through its contents. Most likely, you'll find a bunch of files that you can get rid of.

Delete the Package Files for Apps You've Side-Loaded

If you've side-loaded any apps (as described earlier in this chapter), you can delete their package files. For example, if you placed the package file for Firefox in your Download folder and then installed Firefox from there, the package file will still be there. You can delete it to recover space.

Uninstall Any Apps You Don't Need

To recover space, you can uninstall any apps you've loaded but no longer need.

 You can't uninstall your Kindle Fire's built-in apps.

You can uninstall an app using either the quick way or the slow way. Normally, you'll probably want to use the quick way.

Remove an App the Quick Way

To remove an app the quick way, follow these steps:

1. Tap the Home button to display the Home screen.
2. Tap the Apps button to display the Apps screen.
3. Tap and hold the app you want to remove. The Kindle Fire displays a pop-up menu, as shown here.
4. Tap the Remove From Device button. The Kindle Fire displays a confirmation screen, as shown in Figure 4-29.

FIGURE 4-29   On the confirmation screen for uninstalling an app, tap the OK button.

If the Remove From Device button doesn't appear on the pop-up menu, the app you've chosen is a system app that the Kindle Fire doesn't let you uninstall.

5. Tap the OK button at the bottom of the screen to confirm you want to uninstall the app. The Kindle Fire uninstalls it and then displays the Uninstall Finished screen.
6. Tap the OK button. The Kindle Fire then returns you to the Apps screen, where you can delete other apps if you need to.

 You can also remove an app by tapping and holding it on the Carousel and then tapping the Remove From Device button on the pop-up menu that appears. This method is handy except that if you're removing the app because you don't use it, the app will likely be way back on the Carousel.

Remove an App the Slow Way

Instead of removing an app using the Apps screen or the Carousel, you can remove it using the Settings app. To remove an app this way, follow these steps:

1. Tap the Settings button on the status bar to display the settings bar.
2. Tap the More button on the settings bar to display the Settings screen.
3. Tap the Applications button to display the Applications screen.
4. Tap the Filter By pop-up menu, and then tap the Third Party Applications button. The Applications screen displays only the third-party apps (see Figure 4-30). These are the apps that you can uninstall in this way.
5. Tap the app you want to uninstall. Your Kindle Fire displays the app's screen. Figure 4-31 shows the screen for Ultimate Checklist, an app for keeping checklists.
6. Tap the Uninstall button. Your Kindle Fire displays the Uninstall Application dialog box (shown here) to confirm that you want to uninstall the app.

7. Tap the OK button. Your Kindle Fire uninstalls the app and returns you to the Applications screen in Settings. You can then uninstall other apps if necessary.

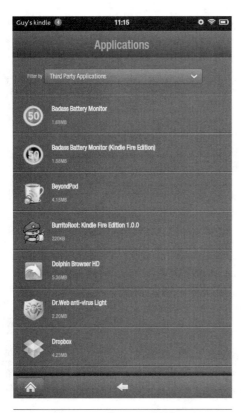

FIGURE 4-30 Choose Third Party Applications in the Filter By pop-up menu to display a full list of the third-party apps on your Kindle Fire.

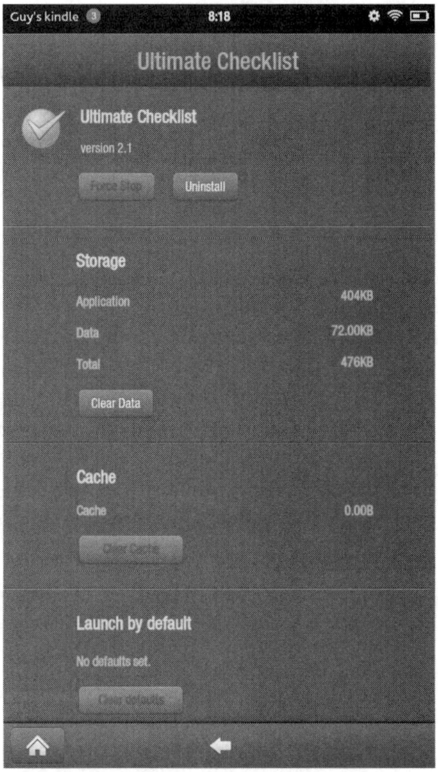

FIGURE 4-31 Tap the Uninstall button on an app's screen to start uninstalling the app.

Project 39: Put Your Photos on Your Kindle Fire

Your Kindle Fire's built-in Gallery app is great for enjoying your photos and sharing them with others. Unlike many other tablets, the Kindle Fire doesn't have a built-in camera, so you'll need to load your photos on your Kindle Fire yourself.

Here are the three easiest ways to put your photos on your Kindle Fire:

- **Copy the photos from your computer** Connect your Kindle Fire to your computer via the USB cable and open a Windows Explorer window or a Finder window to the Pictures folder on the Kindle Fire. You can then copy your photos from your computer and paste them into the Pictures folder. This is the best way of getting a large number of photos onto your Kindle Fire quickly.

- **Load the photos onto Cloud Drive**
  You can upload your photos from
  your computer onto Cloud Drive, and
  then download them from there onto
  your Kindle Fire. At this writing, this
  maneuver is awkward, because you
  must use the Silk browser (or another
  browser) to download the files to your
  Kindle Fire (tap and hold a picture to
  display a dialog box; then tap the Save
  Image button). You can then view
  the picture by tapping the Download
  category in the Gallery app.
- **Use another online-drive service such
  as Dropbox**   If you use Dropbox or a
  similar service, you can use it to store
  your photos and to put them on your
  Kindle Fire. To get a photo from Dropbox
  into the Gallery app on your Kindle Fire,
  follow these steps:
  1. In the Dropbox app, tap the photo's
     name to download it to your Kindle
     Fire.
  2. Tap the photo to open it for viewing.
  3. Tap the Export button at the
     bottom of the screen to display the
     Export This File dialog box (see
     Figure 4-32).
  4. Tap the Gallery button.

FIGURE 4-32   To export a photo
from Dropbox to the Gallery app, tap
the Gallery button in the Export This
File dialog box.

Project 40: Keep Up with Your Social Networks

Your Kindle Fire is a great tool for keeping up with your social networks.

In this project, I'll show you how to get Facebook, LinkedIn, and Twitter set up on
your Kindle Fire so that you can get updates and post your news anywhere you have
an Internet connection.

Use Facebook on Your Kindle Fire

If you have a Facebook account, you'll probably want to use your Kindle Fire to stay
in touch and post your news. You can do so either by using the Kindle Fire's built-in
means of accessing Facebook or by installing a Facebook app and using that.

Understand Your Choices for Accessing Facebook on Your Kindle Fire

If you look on the Apps screen, you'll see a Facebook icon—so you may figure that your Kindle Fire has a Facebook app built in. But if you tap this icon, you'll find it's just a shortcut to the Facebook website: Your default web browser (which is Silk unless you change it) becomes active, and it displays the mobile version of the Facebook website. You can then log in and start using Facebook.

Using the mobile version of the Facebook site is a reasonable approach. Because the mobile version is designed for mobile phones and other devices with small screens, it works pretty well on the Kindle Fire's larger screen. It even has a possible advantage over using a Facebook app, in that it doesn't store data on your Kindle Fire, which might be a security concern if you share your Kindle Fire with your family.

The disadvantage to using the mobile version of the Facebook site is that you don't get all the features of Facebook. If you find you need Facebook features the mobile version doesn't provide, you can side-load the full-on Android Facebook app using the technique described in Project 34, earlier in this chapter. You can also use Silk or another browser to access the desktop version of the Facebook site instead of the mobile version; see the nearby sidebar for details.

DOUBLE GEEKERY

Access the Desktop Version of the Facebook Site in Silk

If you need to use Facebook features the mobile version of the site doesn't offer, try using Silk to access the desktop version of the Facebook site. Follow these steps:

1. Tap the Home button to display the Home screen.
2. Tap the Web button to launch or switch to Silk.
3. Tap the address box and type in **facebook.com**, or as much is needed to identify the site.
4. In the list of matches, tap the button for the site.

If Silk displays the mobile version of the site, this is because you've set Silk to use the mobile version of sites (as recommended in Project 28 for general use). In this case, you'll need to set it to use the desktop version. Follow these steps to switch back:

1. In Silk, tap the Settings button to display the Settings screen.
2. Scroll down to the bottom of the screen.

3. Tap the Desktop Or Mobile View button to display the Desktop Or Mobile View dialog box (shown here).
4. Tap the Desktop: Optimize For Desktop View option button.

Now tap the Refresh button at the right end of the address box to reload the Facebook page. This time, you'll see the desktop version rather than the mobile version.

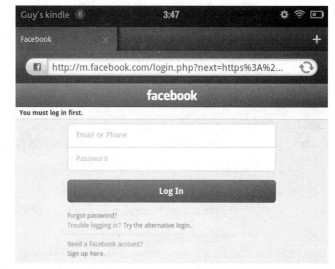

Access the Mobile Version of Facebook Using Silk

To access the mobile version of Facebook using the Silk web browser, simply tap the Facebook icon on the Home screen. (If the Facebook icon doesn't appear on the Home screen, tap the Apps button, and then tap the Facebook icon on the Apps screen.)

On the Facebook login screen, shown here, type your user name and password, and then tap the Log In button.

In the Confirm dialog box, shown here, choose whether you want Silk to remember your password. Tap the Not Now button, the Remember button, or the Never button, as appropriate.

You then see the Welcome To Facebook page for the mobile site, and you can navigate it as usual.

Install and Use the Facebook for Android App

If you want to use Facebook extensively on your Kindle Fire, install the Facebook for Android app. This app is free but not available in the Appstore, so you'll need to get it from another source and side-load it. Follow these general steps:

1. Get the package file for Facebook for Android using one of the methods discussed in Project 33: extracting the package file from an Android device that can access Android Market or downloading the package file from a website or other online resource.
2. Use the technique explained in Project 34 to side-load Facebook for Android on your Kindle Fire.
3. After installing Facebook for Android, tap the Open button to launch it.

When you run the Facebook app for the first time, it displays the Login screen (shown here). Type your e-mail address and password, and then tap the Login button.

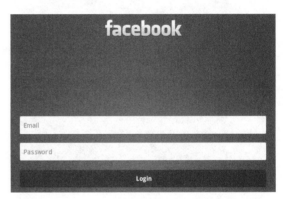

Once you get into the Facebook app, it's easy to use. These are the main moves you'll need:

- Tap the Facebook logo at the top of any screen to display the home screen.
- Tap the button in the upper-left corner of the home screen to display a screen of Favorites, Lists, and Apps that you can quickly jump to. Figure 4-33 shows the Favorites and Lists; the Apps section appears further down the screen.
- To choose settings for the Facebook app, tap the Menu button at the bottom of the screen, and then tap the Settings button. On the Settings screen (see Figure 4-34), you can change the refresh interval (from 30 Minutes to 4 Hours—or Never) and choose whether to receive notifications. If you choose to receive notifications, specify which items to notify you about and which forms of notification to use (vibration or a ringtone; the Phone LED option doesn't work because the Kindle Fire doesn't have an LED). Tap the Back button when you finish choosing settings.

- To refresh a screen with the latest data, tap the Menu button at the bottom of the screen, and then tap the Refresh button on the menu panel.

 To log out of Facebook *and remove your Facebook data from your Kindle Fire,* tap the Menu button from the home screen, and then tap the Logout button. Tap the Yes button in the Logout dialog box that appears.

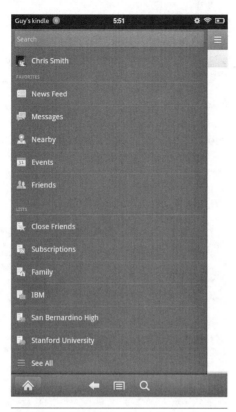

FIGURE 4-33 Tap the button at the upper-left corner of the Facebook app to quickly access any of the main areas of Facebook.

FIGURE 4-34 On the Settings screen for the Facebook app, you can choose which notifications to receive, and even set a ringtone for Facebook notifications.

Use LinkedIn on Your Kindle Fire

At this writing, LinkedIn is the leading social networking app for business purposes. If you have a LinkedIn account, you'll likely want to access LinkedIn from your Kindle Fire to stay in touch with your contacts and keep up to date.

One way to access LinkedIn is by opening the Silk browser and going to the LinkedIn website. You can display either the mobile version or the desktop version of the site, depending on whether you've chosen the Desktop: Optimize For Desktop View option button or the Mobile: Optimize For Mobile View option button on the Settings screen for Silk.

Using the browser works okay, but what you'll normally want to do is install the LinkedIn app on your Kindle Fire and use that instead. In this section, I'll show you how to get the LinkedIn app up and running and how to access your account.

Install the LinkedIn App on Your Kindle Fire

Unlike the Facebook app, the LinkedIn app for Android is available on the Appstore—and it's free. So you can get the LinkedIn app by going to the Appstore, searching for **linkedin**, and then tapping the LinkedIn result. Tap the Free button, and then tap the Get App button as usual.

Log In and Start Using LinkedIn on Your Kindle Fire

When your Kindle Fire finishes installing the LinkedIn app, tap the Open button. You'll then see the Sign In screen (shown here).

Type your e-mail address and password, and then tap the Sign In button. The LinkedIn app then displays the Sync LinkedIn Contacts dialog box, shown here, asking if you want to sync your LinkedIn contacts with your existing contacts. Tap the Sync All option button, the Sync With Existing Contacts option button, or the Do Not Sync LinkedIn Contacts option button, as appropriate.

After you sign in, LinkedIn displays the home page, the top of which appears in Figure 4-35. From here, you can quickly search using the Search box or access one of the main areas of LinkedIn by tapping its button at the top of the screen. For example, tap the You button to display your profile, or tap the Inbox button to display your inbox.

 To refresh the information displayed, tap the Menu button to display the Menu panel, and then tap the Refresh button.

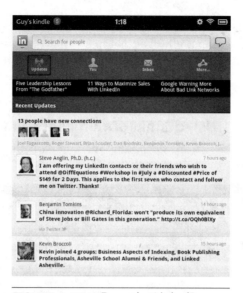

FIGURE 4-35 From the LinkedIn home screen, you can quickly view your updates (shown here), your profile, your inbox, or other items.

Use Twitter on Your Kindle Fire

If you want to keep up with the latest happenings, and share your own, you'll almost certainly want to use Twitter on your Kindle Fire.

Your Kindle Fire doesn't come with a Twitter app, so you have a choice of installing a free app or simply using the Silk browser to access Twitter. You can display either the mobile version or the desktop version of the site, depending on whether you've chosen the Desktop: Optimize For Desktop View option button or the Mobile: Optimize For Mobile View option button on the Settings screen for Silk.

If you're a serious user of Twitter, you'll probably want to install and use the Twitter app. This section shows you how to get the app and get started with it.

Install the Twitter App on Your Kindle Fire

Amazon makes a Twitter app available for free on the Appstore. To get the app, tap the Apps button on the Home screen, and then tap the Store button to access the Appstore. Type **twitter** into the Search box, and then tap the Twitter result. Tap the Free button, and then tap the Get App button to start the installation.

Log In and Start Using Twitter on Your Kindle Fire

When your Kindle Fire has installed Twitter, tap the Open button to open the app.

The first time you run the Twitter app, it displays the Welcome To Twitter screen. Tap the Sign In button to display the Sign In screen (shown here), type your user name (or your e-mail address) and password, and then tap the Sign In button.

Once you've signed in, the Twitter app displays the home screen (the top part of which is shown in Figure 4-36), on which you can read the latest tweets. From here, you can quickly access the other main areas of Twitter. For example:

- Tap the Compose button in the upper-right corner of the screen at the top to display the screen for creating a tweet (see Figure 4-37). You can then type a text tweet, add an existing photo by tapping the Add Photo button, add people by tapping the Add People button, or add your location by tapping the Add Location button.

FIGURE 4-36   On the Twitter home screen, you can quickly catch up with the latest tweets.

- Tap the Connect button to connect to other people.
- Tap the Discover button to search by a hashtag or keyword.
- Tap the Me button to review or update your profile. To change your profile photo, tap the Edit Profile button, and then tap the Change Profile Photo button on the Edit Profile screen.

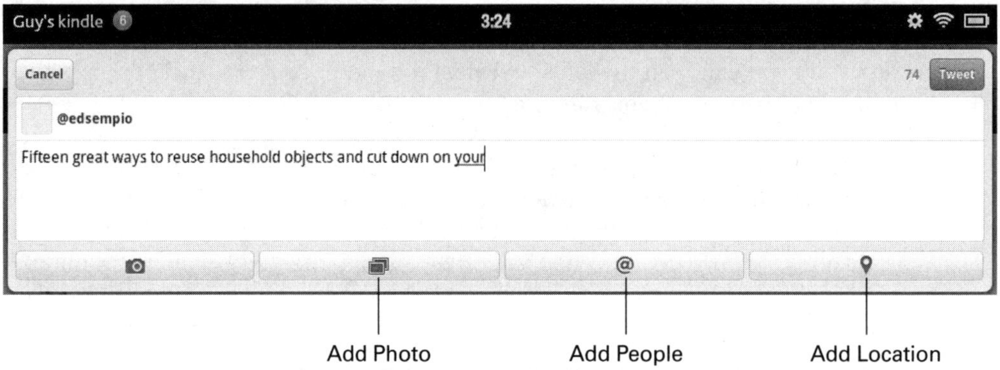

FIGURE 4-37 On the screen for creating a tweet, you can type a text tweet, add an existing photo, add people, or add your location.

5   Kindle Fire at Work Geekery

As you can see from its built-in apps, Amazon designed the Kindle Fire for pleasure rather than work. But as you'll see in this chapter, not only is there nothing to stop you from using your Kindle Fire for work, but your Kindle Fire can be a great tool for getting your work done when you're out and about.

In this chapter, I'll show you first how to connect your Kindle Fire to the wireless network at your workplace, installing digital certificates if necessary for the connection. We'll then look at how to connect your Kindle Fire to Exchange Server systems and how to manage your e-mail like a pro before moving on to dig into ways to get your contacts, calendars, tasks, and notes onto your Kindle Fire. We'll finish by adding a way to send text messages from your Kindle Fire, installing a suite for creating Office documents on it, and giving yourself access to your essential files no matter where you are.

Project 41: Connect Your Kindle Fire to the Wireless Network at Your Workplace

In this project, I'll show you how to connect your Kindle Fire to the wireless network at your workplace. The process is similar to connecting to your wireless network at home or connecting to a wireless hotspot, but you'll normally need to use more sophisticated means of authentication than a password—for example, a digital certificate. In this case, you must install the digital certificate on your Kindle Fire before you can establish the wireless connection.

First, find out which credentials you need for the wireless network: password, user name and shared secret, digital certificate, or other credentials. If you need to use a digital certificate, install it as explained in the first subsection. Then connect to the wireless network as described in the second subsection.

DOUBLE GEEKERY

Connect Your Kindle Fire to Your Work Network Across the Internet Using a VPN

The best way to connect your Kindle Fire to your work network across the Internet is by using a virtual private network, or VPN. A VPN uses encryption to create a secure tunnel across the Internet from one computer to another—for example, from your Kindle Fire to the VPN server on your work network.

Although the Android OS has VPN capability built in, the Kindle Fire's user interface doesn't let you reach this feature. So, unless you root your Kindle Fire and install a modified version of the operating system that lets you reach the VPN feature, the only way to connect your Kindle Fire via a VPN is to set up a VPN using your computer or your router, and then connect the Kindle Fire through that.

To connect the Kindle Fire through your computer, follow these general steps:

1. Connect your computer to your router via an Ethernet cable, not via wireless.
2. Set up a VPN connection from Network And Sharing Center (on Windows) or the Network preferences pane (on the Mac).
3. Set up the Internet Connection Sharing feature in Windows or the Internet Sharing feature on the Mac.
4. Connect your Kindle Fire to your computer via the wireless network. The following illustration shows a Kindle Fire connected via a VPN from a computer.

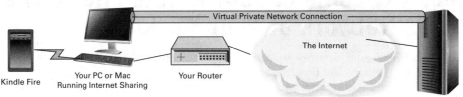

To connect the Kindle Fire through your router, go into your router's configuration screens and set up the VPN. Most routers support web-based configuration, so you can access them by typing the appropriate IP address in your web browser—for example, **10.0.0.2** or **192.168.1.1**. Beyond that, you're on your own. All routers are different, and you may need to consult your router's documentation to set up the VPN connection. The following illustration shows a Kindle Fire connected via a VPN from the router.

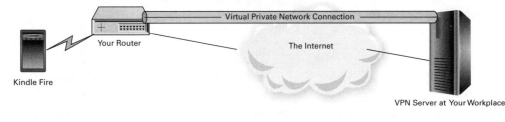

Install a Digital Certificate on Your Kindle Fire

To install a digital certificate on your Kindle Fire, follow these steps:

1. Get the certificate in a suitable format (see the upcoming sidebar) from your network administrator or from your computer:
   - **Windows**   Click the Start button to open the Start menu, then type **certmgr .msc** in the Search box, and press ENTER to open the Certificate Store. Find the certificate, right-click it, and choose All Tasks | Export to launch the Certificate Export Wizard. On the Export File Format screen, select Base-64 Encoded X.509.
   - **Mac**   Click the Desktop, choose Go | Utilities, and then double-click the Keychain Access icon in the Utilities window. Locate the certificate, CTRL-click or right-click it, and then click Export on the context menu to start the process of exporting the certificate. In the Save As dialog box, choose either Certificate (.cer) or Personal Information Exchange (.p12) in the Save As pop-up menu.
2. Connect your Kindle Fire to your PC or Mac, and then copy the certificate to the Download folder on your Kindle Fire.

 You can also attach the certificate to an e-mail message and send it to an e-mail account on your Kindle Fire. From there, save the certificate to your Download folder, and then follow the list from Step 4 onward.

3. Tap the Disconnect button on the You Can Now Transfer Files From Your Computer To Kindle screen to disconnect your Kindle Fire from your computer.
4. Tap the Settings button on the status bar to display the settings bar.
5. Tap the More icon to display the Settings screen.
6. Tap the Security button to display the Security screen (shown here).

![Security settings screen on Kindle Fire showing Guy's kindle, time 1:07, Lock Screen Password toggle set to OFF, Credential Storage, and Device Administrators (Add or remove device administrators).]

7. Tap the Credential Storage button to display the Credential Storage screen (shown here).

8. Tap the Set Credential Storage Password button to display the Credential Storage Password screen (shown here).

9. Type a password of at least eight characters in the Enter Password box and the Confirm Password box.
10. Tap the Finish button to return to the Credential Storage screen. You'll see the Use Secure Credentials Switch has moved to the On position because you set a password. Leave it set to On.
11. Tap the Install Secure Credentials button. Your Kindle Fire searches for the certificate and then displays the Name Certificate dialog box (shown here) containing its name and details.

Name Certificate

Certificate name:

Network Certificate

The package contains:
one user certificate

OK   Cancel

12. Make sure the certificate is the correct one, and then tap the OK button.
13. If your Kindle Fire prompts you to enter a password to extract the certificate from a PKCS12 file, as shown here, type the password, and then tap the OK button.

Extract from Wi-Fi Certificate 1.p12

Enter the password to extract the certificates.

OK   Cancel

DOUBLE GEEKERY

Get Your Certificate in a Suitable Format for Your Kindle Fire

Ideally, your network administrator will provide you with a digital certificate to put on your Kindle Fire for authenticating it to the wireless network. But there's a fair chance you'll need to sort out the certificate yourself—so here are the technical details you need:

- The certificate must be in either Base-64 Encoded X.509 format or PKCS12.
- If the certificate is in the X509 format, it needs to have the .crt file extension rather than the .cer file extension for your Kindle Fire to use it. If you export an X.509 certificate from the Certificate Store in Windows, Windows creates a file with a .cer file extension. To make this file work on your Kindle Fire, you'll need to change the file extension to .crt.
- If the certificate is in the PKCS12 format, it needs the .p12 file extension.

---

14. Your Kindle Fire installs the certificate, and then displays a confirmation dialog box (shown here).

Network Certificate is installed.

Dismiss

15. Tap the Dismiss button to dismiss the dialog box.

Connect to the Wireless Network

Armed with your credentials, you're ready to connect to the wireless network at your workplace. Follow these steps:

1. Tap the Settings icon on the status bar to display the settings bar.
2. Tap the Wi-Fi icon to display the Wi-Fi panel.
3. Tap the network's name to display the dialog box for connecting to the network (see Figure 5-1).
4. Use the pop-up menus to specify the means of authentication. For example, if the network uses EAP security, open the EAP Method pop-up menu and tap the PEAP option button, the TLS option button, or the TTLS option button, as needed. See the next sidebar titled "Understand the Acronyms for Authentication" for an explanation of what the acronyms mean.

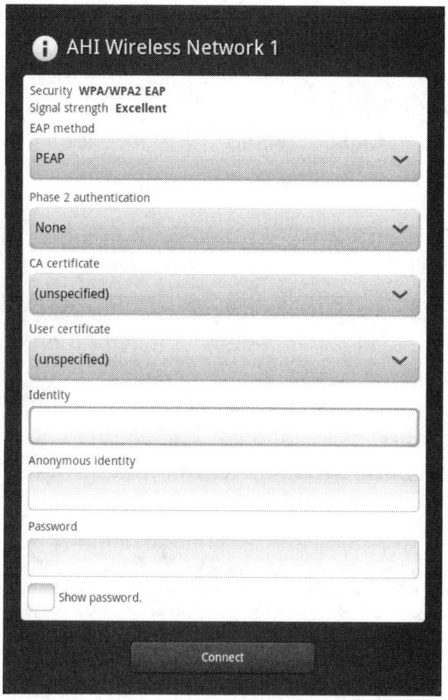

FIGURE 5-1   In this dialog box, enter the means of authentication for the wireless network, and then tap the Connect button.

DOUBLE GEEKERY

Understand the Acronyms for Authentication

Authentication uses a barrage of protocols that look like alphabet soup. A *protocol* is a set means of communication between computers. Here are the terms you may need to know to make sense of authentication:

- **EAP**   Extensible Authentication Protocol is a family of protocols for authenticating computers on networks.
- **PEAP**   Protected Extensible Authentication Protocol is a protocol for authenticating computers on wireless networks without having to use certificates.
- **TLS**   Transport Layer Security is a protocol for communicating securely across an insecure network (such as the Internet).
- **TTLS**   Tunneled Transport Layer Security is another protocol for authenticating computers on wireless networks without using certificates.
- **PAP**   Password Authentication Protocol is pretty much what its name says. PAP transmits the password unencrypted, so it's not secure—but some companies and organizations still use it.

- **MSCHAP** Microsoft Challenge-Handshake Authentication Protocol is a protocol for authenticating using passwords.
- **MSCHAPV2** Microsoft Challenge-Handshake Authentication Protocol version 2 is the second version of MSCHAP.
- **GTC** Generic Token Card is a protocol for authenticating by using a security token or a one-time password.

---

5. Use the CA Certificate pop-up menu and the User Certificate pop-up menu, as needed, to choose the certificates you installed for the network connection.
6. Type your user name in the Identity box.
7. Type your password in the Password box. As usual, you can select the Show Password check box to reveal what you've typed and make sure it's correct.
8. Tap the Connect button to connect to the network.

Project 42: Connect Your Kindle Fire to Exchange Server

If you use your Kindle Fire at work, you may need to connect it to an Exchange Server system so that you can access your e-mail, contacts, and schedule information.

Unlike standard Android phones and tablets, the Kindle Fire doesn't include the Exchange ActiveSync connector that enables you to connect to Exchange Server using the Email app. But you can connect to Exchange Server by using a third-party solution called TouchDown, as you'll see in this project.

TouchDown comes in two versions: a free, ad-supported version called Exchange By TouchDown that gives you a 30-day free trial, and a pay version called Exchange By TouchDown Key that costs $19.99.

Download, Install, and Set Up TouchDown

Download and install TouchDown from the Appstore:

1. From the Home screen, tap the Apps button to display the Apps screen.
2. Tap the Store button to display the Store screen.
3. Tap the Search box, type **nitrodesk**, and then tap the Search button.
4. Tap the NitroDesk TouchDown search result.
5. On the screen of search results, tap the Free button on the Exchange By TouchDown button, and then tap the Get App button.

6. When the installation finishes, tap the Open button to open Exchange By TouchDown. You'll then see the Welcome screen (shown here).

Connect TouchDown to Your Exchange Server

To connect TouchDown to your Exchange server, follow these steps:

1. Tap the Configure Your Account button to display the Configure Your Account screen (shown here).

2. Type your e-mail address in the Email Address box and your Exchange password in the Password box.

3. Tap the Try AutoDiscovery button. TouchDown contacts the Exchange servers at the address you gave and tries to discover which server to use. If AutoDiscovery works, go to the end of this list.

4. If AutoDiscovery can't identify the server after several minutes' trying, TouchDown displays a message saying that AutoDiscovery may not be supported for your server. Tap the Next button to display the Connection Details screen (shown here).

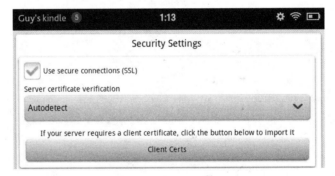

5. Type in the full details of your Exchange account, and then tap the Next button. TouchDown displays the Security Settings screen (shown here).

6. Leave the Use Secure Connections check box selected and the Server Certificate Validation drop-down list set to Autodetect.
7. Tap the Next button. TouchDown displays the Protocols To Check For screen (shown here).

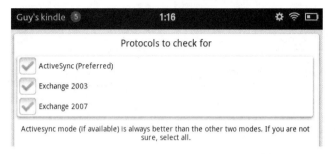

8. Choose which protocols to use for syncing Exchange. When you're setting up TouchDown for the first time, it's best to leave all three protocols selected. Afterward, you can try using only ActiveSync, which is the preferred protocol.
9. Tap the Next button. TouchDown displays the Start Configuration? screen, which warns you that the sync may take a long time.
10. Tap the Next button. TouchDown contacts the server and establishes the connection. TouchDown then displays the Configuration Progress screen (shown here).

11. Click the Close button (not shown in the previous illustration).

Your inbox then appears (see Figure 5-2), and you can start working with your Exchange items.

FIGURE 5-2 After TouchDown connects to the Exchange server, you can start working with your e-mail messages, tasks, contacts, and calendar.

Project 43: Manage Your E-mail Like a Pro

Your Kindle Fire is a great device for keeping up with your e-mail on the move. Its screen is big enough to make messages easy to read, and the on-screen keyboard is adequate for writing short or medium-length messages.

In this project, I'll show you how to manage your e-mail like a pro on your Kindle Fire. First, we'll get your Kindle Fire set up with the accounts you need. Then we'll look at how to manage your messages by using batch-editing, flagging, drafts, and other expert techniques.

Set Up Your Kindle Fire with the E-mail Accounts You Need

Chances are you set up an e-mail account while setting up your Kindle Fire—but do you use only a single e-mail account?

In this section, you'll learn how to add and remove accounts and how to configure your accounts to work the way you prefer.

Set Up an E-mail Account on Your Kindle Fire

To set up an e-mail account, follow these steps:

1. Open the Email app if it's not already open. For example, tap the Apps button on the Home screen, and then tap the Email icon.
2. If the Email app opens to within a particular account (for example, an inbox), tap the Menu button, and then tap the Accounts button on the Menu panel to display the Accounts screen.
3. Tap the Menu button to display the Menu panel, and then tap the Add Account button to display the Select E-mail Provider screen (shown here).

4. Tap the button for the e-mail provider. The Email app displays the Sign-In screen (shown here). This example uses Gmail, but the screens for the other e-mail providers are similar. For the Other category, you typically need to enter the details of your account's e-mail servers.

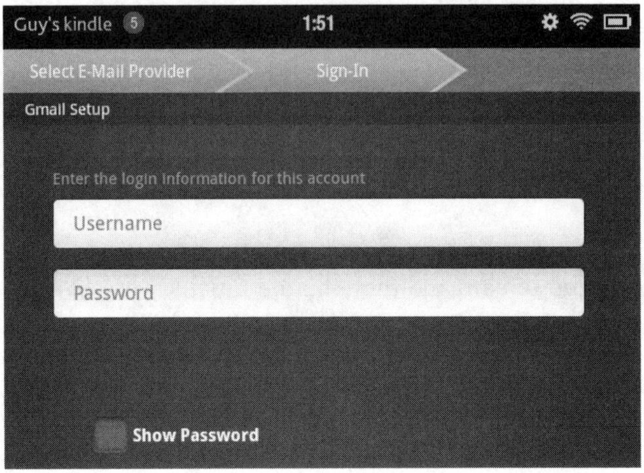

5. Type your e-mail address in the Username box and your password in the Password box. Select the Show Password check box if you want to be able to see the characters in your password.

6. Tap the Next button at the bottom of the screen (not shown in the previous illustration). The Email app connects to the mail servers and verifies your account. It then displays the Finish screen (shown here for Gmail).

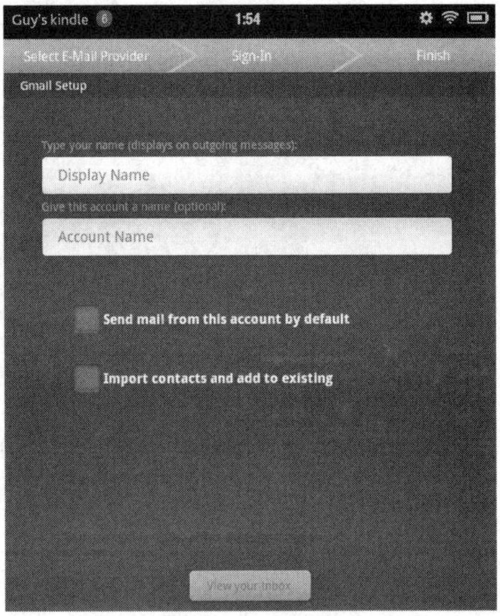

7. Type your name in the Display Name box the way you want it to appear on messages you send.

8. Type a descriptive name for the e-mail account in the Account Name box, so you'll be able to distinguish it from your other accounts.

9. Select the Send Mail From This Account By Default check box if you want to use this account as your default account.

10. Select the Import Contacts And Add To Existing check box if you want to import this account's contacts and add them to the contacts already in the Contacts app.

11. Tap the View Your Inbox button to finish setting up the account and to display your Inbox.

12. If you selected the Import Contacts And Add To Existing check box, the Email app displays the Duplicates May Occur dialog box (shown here), warning you that importing your contacts may create duplicates and that it will not overwrite existing contacts. Tap the Import button to continue.

13. If the Email app displays the Error Importing E-mail Contacts dialog box (shown here), telling you that you need to log in to your e-mail account in a separate browser window, tap the Continue button, and then use the browser to log into your account so that the Email app can access your contacts.

Duplicates May Occur

Imported contacts may create duplicate entries (contacts will not be overwritten).

Are you sure you want to import?

Import   Skip Import

Error Importing E-Mail Contacts

To import your gmail.com contacts, you must login to gmail.com in a separate browser window. Do you wish to continue?

Continue   Skip Import

Remove an E-mail Account from Your Kindle Fire

Removing an e-mail account from your Kindle Fire is easy once you know how—but it's not immediately obvious to most people. Follow these steps:

1. Open the Email app if it's not already open.
2. If the Email app opens to within a particular account (for example, an inbox), tap the Menu button, and then tap the Accounts button on the Menu panel to display the Accounts screen.
3. Tap and hold the button for the account until the Account Options dialog box (shown here) opens.

Account options

Open

Check mail

Empty trash

Account settings

Import contacts

Remove account

4. Tap the Remove Account button. The Email app displays the Remove Account dialog box (shown here).

Remove Account

The account "Gmail" will be removed from Email.

OK   Cancel

5. Tap the OK button. The Email app removes the account.

Configure Your E-mail Accounts to Work the Way You Prefer

To make your e-mail accounts work the way you prefer, spend a few minutes configuring each of them. This section explains the settings you can customize to change how the account behaves.

First, open the Settings screen for the account. Follow these steps:

1. In the Email app, tap the account's button on the Accounts screen to display the account's inbox.
2. Tap the Menu button to display the Menu panel.
3. Tap the Settings button to display the Settings screen. Figure 5-3 shows the upper part of the Settings screen. Figure 5-4 shows the lower part.

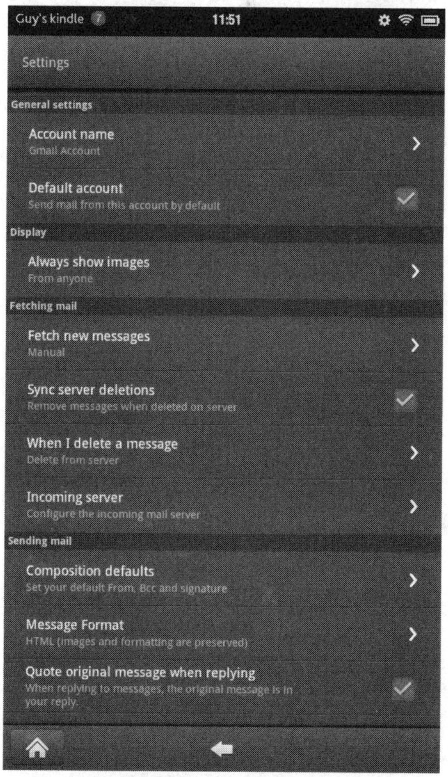

FIGURE 5-3 The upper part of the Settings screen for an e-mail account contains the General settings, the Display settings, the Fetching Mail settings, and the Sending Mail settings.

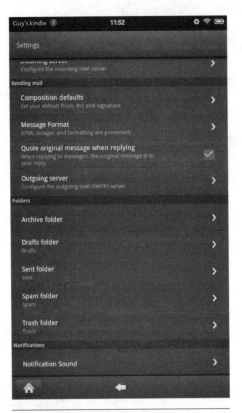

FIGURE 5-4 The lower part of the Settings screen for an e-mail account contains the Folders settings and the Notifications settings.

In the General Settings section at the top of the Settings screen, you can set these options:

- **Account Name** This button shows the name you've given the account—for example, Gmail Account or Work Hotmail. To change the name, tap this button, type the new name in the Account Name dialog box, and then tap the OK button.
- **Default Account** Select this check box to make this account your default account for sending messages. Selecting this check box for one account clears it for whichever other account it was previously selected.

Choose the From Contacts Option for the Always Show Images Setting

These days, it's hard to avoid receiving at least some spam—unwanted messages. When you do, you can simply delete them. But malefactors have another trick up their sleeve: remote images, which are also called *web bugs*. By including, in a message, a reference to an image stored on a remote server, a spammer can learn not only when you open that message but also your IP address and approximate geographical location.

To avoid this problem, choose the From Contacts option for the Always Show Images setting, not the From Anyone option. This tells the Email app not to load remote images from people you don't know. The images then appear as placeholders in your messages. You can then look through a message to see whether it's safe to load the images. If it is, tap a placeholder to display its image.

In the Display section, you can set just one option:

- **Always Show Images**   To choose whether the Email app displays remote images included in messages, tap this button, and then tap the No option button, the From Contacts option button, or the From Anyone option button in the Always Show Images dialog box (shown here). See the nearby sidebar for advice.

Always show images

No

From contacts ●

From anyone

Cancel

In the Fetching Mail section, you can choose these settings:

- **Fetch New Messages** Tap this button to display the Push Folders dialog box (shown here), and then tap the Push option button or the Manual option button, as needed. Choosing Push makes the server "push" the messages out to your Kindle Fire, so you get them automatically rather than having to check manually. Only some mail servers support Push.

Push folders

Push

Manual

Cancel

- **Sync Server Deletions** Select this check box to make the Email app remove messages from your inbox and other folders when you delete them on the mail server using another computer or device.
- **When I Delete A Message** Tap this button to display the When I Delete A Message dialog box (shown here), and then tap the Do Not Delete On Server option button, the Delete From Server option button, or the Mark As Read On Server option button, as needed. Which of these you find most useful depends on how you use your Kindle Fire for e-mail. If you're using it to manage your e-mail, choose the Delete On Server option button.

When I delete a message

Do not delete on server

Delete from server

Mark as read on server

Cancel

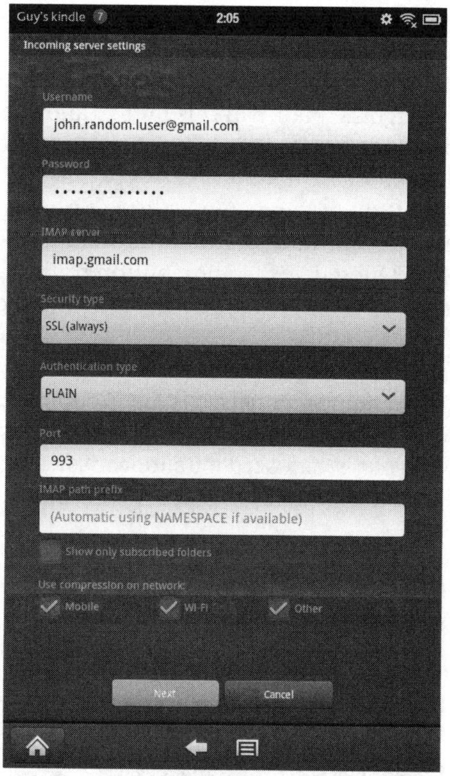

FIGURE 5-5   Use the Incoming Server
Settings screen to configure the incoming
mail server for your e-mail account.

- **Incoming Server**   If you need to change the settings for your incoming mail
  server, tap this button, and then work on the Incoming Server Settings screen
  (see Figure 5-5). Normally, the Email app picks up these settings automatically
  when you set up the account, but you may sometimes need to change some of them.
  For example, you may need to change the mail server name, the authentication type,
  or the security type. Tap the Next button when you're ready to test the settings.

In the Sending Mail section, you can choose these settings:

- **Composition Defaults** To change your display name, your signature, or other options for outgoing mail, tap this button and then work on the Message Composition Options screen (see Figure 5-6). Select the Use Signature check box if you want to include a signature, and then type the signature in the Signature box, tapping the ENTER button to create a new line as needed. In the Signature Position area, specify where to put the signature on replies and forwarded messages by selecting the Before Quoted Text option button or the After Quoted Text option button. Tap the Save button when you finish making your changes.

 On the Message Composition Options screen, you can type an e-mail address in the Bcc All Messages To box if you need the Email app to automatically send a copy of each outgoing message to a particular account. Normally, you don't need to do this, especially if your e-mail account automatically stores your sent messages in a Sent folder.

FIGURE 5-6 On the Message Composition Options screen, set your display name and signature for outgoing messages.

- **Message Format**   To control whether the Email app sends plain-text messages or HTML messages (with pictures and formatting), tap this button. In the Message Format dialog box (shown here), tap the Plain Text option button or the HTML option button as needed.

- **Quote Original Message When Replying**   Select this check box if you want the Email app to include the original message when you create a reply. Clear this check box if you want to send just what you type in the reply. If you clear this check box, your reply displays a Quote Message button that you can tap to insert the original message when you need it.
- **Outgoing Server**   If you need to change the settings for your outgoing mail server, tap this button, and then work on the Outgoing Server Settings screen (see Figure 5-7). Normally, the Email app picks up these settings automatically when you set up the account, but you may sometimes need to change the configuration. When you finish making your changes, tap the Next button to make the Email app check the settings; if they don't work, you'll get an error dialog box pointing out the problem.

In the Folders section of the Settings screen, you can set the folders the Email app is using for the account. The selection of folders varies depending on the e-mail account, but most accounts include a Drafts folder, a Sent folder, a Spam folder, and a Trash folder. To change a folder, tap its button, and then tap the appropriate option button in the dialog box that opens. Figure 5-8 shows the Drafts Folder dialog box, which you use to specify the folder to use as the Drafts folder.

In the Notifications area of the Settings screen, you can tap the Notification Sound button to display the Notification Sound dialog box, and then tap the option button for the sound you want to hear as your aural notification—for example, Caffeinated Rattlesnake or Kzurb Sonar. Tap the OK button when you've made your choice.

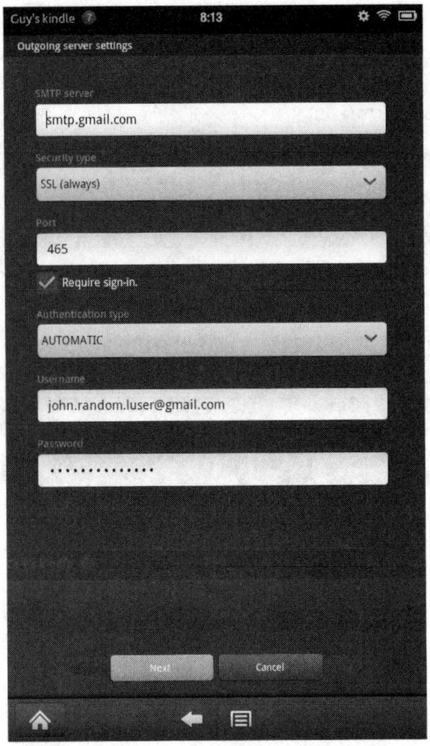

FIGURE 5-7 You can change the settings for the account's outgoing mail server on the Outgoing Server Settings screen. For example, you may need to change the authentication type.

Manage Your E-mail Messages

I'm sure you know the basics of replying to messages, forwarding them to other people, and using the Move command on the Menu panel to move them to suitable folders for safe keeping.

In this section, we'll look at how to manage your e-mail messages, doing everything from switching among your folders, managing messages from your inbox, and batch-editing messages, to saving drafts and using the Bcc field effectively.

Switch Among the Folders in Your Account

When you tap an account's name on the Accounts screen, the Email app displays your inbox for that account. To display the contents of a different folder, follow these steps:

1. Tap the Menu button to display the Menu panel.
2. Tap the Folders button to display the Folders list.
3. Tap the folder you want to display.

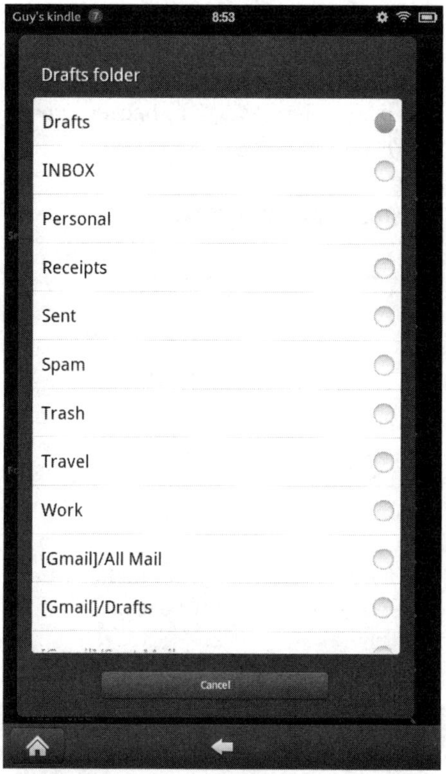

FIGURE 5-8   Tap the option button for
the folder you want to use for a particular
function—in this case, for drafts.

To go back to the inbox, tap the Menu button, tap the Accounts button, and then
tap the account's name on the Accounts screen.

Manage Messages from the Inbox

To get through your messages quickly, cull any worthless messages straight from your
inbox without opening them. To delete a message, tap its button and hold until the
Actions dialog box (shown here) opens, and then tap the Delete button.

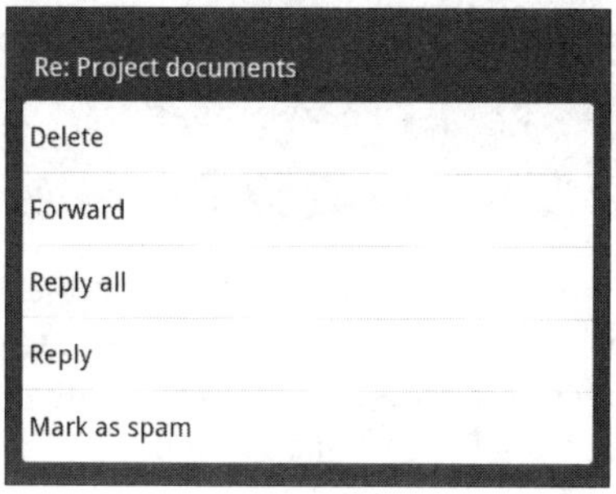

From the Actions dialog box, you can also forward a message, reply to it, or reply to all recipients. Normally, it's best to open a message and read it in full before you take any of these actions. The other action that you can usefully take from the Actions dialog box is Mark As Spam, because you'll often be able to tell that a message is spam without opening it.

Batch-Edit Your E-mail Messages

Instead of dealing with the e-mail messages in an inbox or folder one by one, you can use batch-editing to manipulate multiple messages at once. To use batch-editing, follow these steps:

1. Open the inbox or folder that contains the messages.
2. Tap the Menu button to display the Menu panel, and then tap the Edit List button. The Email app displays a check box to the left of each message's button and shows command buttons at the bottom of the screen. Figure 5-9 shows an example using Gmail.

 Tap the check box to the left of the Mark As Read button at the bottom of the screen to select or clear all the check boxes.

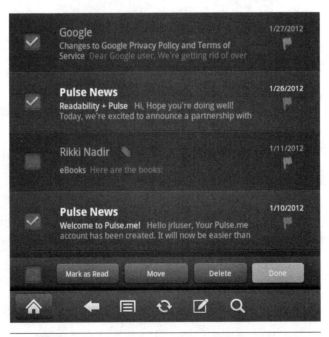

FIGURE 5-9 Tap the Menu button and then tap the Edit List button to turn on Edit mode. You can then select the check box for each message you want to affect, and then tap the appropriate command button.

3. Select the check box for each message you want to affect.
4. Tap the appropriate command button. For example, tap the Move button to display the Mailboxes screen, and then tap the mailbox to which you want to move the messages.

Mark a Message as Unread

When you receive a new message, your inbox displays its sender, subject, and date in boldface to indicate you haven't yet read it. When you open the message and then return to your inbox, the sender, subject, and date appear dimmed, so you can tell instantly which messages you've read and which you haven't.

When you're triaging your e-mail, you may want to look quickly at a message but then mark it as unread so that you can see it still needs your attention. To mark the message as unread, tap the Menu button to display the Menu panel, and then tap the Edit List button. The Email app displays a check box to the left of each message's button and shows command buttons at the bottom of the screen. Select the check box for each read message you want to mark as unread, and then tap the Mark As Unread button at the bottom of the screen.

 You can use a similar technique to mark unread messages as read: Tap the Menu button, tap the Edit List button, select the check boxes for the read messages, and then tap the Mark As Read button. But usually marking read messages as unread is more useful than marking unread messages as read.

Flag Each Message That Has Long-Term Importance

An easy way to keep track of the messages you need to deal with is to flag them. To flag a message, tap the flag icon on the right side of its button in the inbox or a folder. If the message is open, tap the down-arrow button to the right of its subject line, and then tap the flag icon to the right of the address, as shown here.

The flag persists until you remove it by tapping the flag icon again. You can use flagging for whatever purpose you choose, but its basic advantage over marking the message as unread is that the flag stays in place when you open the message for reading.

Empty the Trash for an E-mail Account

When you delete a message, the Email app puts it in the Trash rather than actually getting rid of it. You can retrieve messages from the Trash until you empty it. You can retrieve a message in either of two ways:

- Open the message and then use the Move command to move it to a different folder.
- Tap and hold the message's button until the Actions dialog box opens, and then tap the Undelete button.

When you're ready to empty the Trash, tap and hold the account's button on the Account screen until the Account Options dialog box appears; then tap the Empty Trash button. There's no confirmation.

Change the Account You're Sending a Message From

If you start an e-mail message from the wrong account, you don't need to scrap the message and start again. Just tap the Send As pop-up menu below the Message Text area to display the pop-up menu of your accounts, and then tap the account you want to send the message from.

Save a Message as a Draft So You Can Finish It Later

When you don't have time to complete an e-mail message you've started writing, tap the Save Draft button at the bottom of the screen to save the message as a draft so that you can finish it later.
To continue the message, follow these steps:

1. Tap the Menu button to display the Menu panel.
2. Tap the Folders button to display the Folders screen.
3. Tap the Drafts button to display the Drafts screen.
4. Tap the message to open it.

 For reasons as yet unexplained, when you tap the Save Draft button, some Gmail accounts not only save a draft *but also send a copy to the recipient*. At this writing, the only way to avoid this happening is not to enter the recipient's address until you're ready to send the message.

Send a Message to a Group Without Revealing the E-mail Addresses

When you need to send an e-mail message to a group of people who don't necessarily know each other, don't put all the e-mail addresses in the To box or the Cc box, because each recipient will be able to see all the other addresses.

Instead, put your own address in the To box, and then tap the Cc/Bcc button area to display the Cc field and Bcc field. Put each e-mail address in the Bcc field, and each recipient will see only his or her own e-mail address, not those of the other Bcc recipients. (They'll also see your e-mail address as the recipient as well as the sender.)

Project 44: Load All Your Contacts on Your Kindle Fire

Your Kindle Fire comes with an easy-to-use Contacts app that you can use to keep in touch with your family, friends, and business contacts. You can create contact records directly on your Kindle Fire when you need to, but you'll have to type in each item of information. So what you'll normally want to do is import your contacts from whichever contact-management application you use on your computer.

In this project, I'll show you how to load your contacts from four widely used sources:

- Windows Contacts
- Microsoft Outlook
- OS X Address Book and Contacts
- Google Contacts

 OS X versions up to and including Lion (OS X 10.7) call the contact-management application Address Book. In Mountain Lion (OS X 10.8), the application is called Contacts, matching the Contacts app on the iPhone and iPad. Despite the name change, the application works in almost exactly the same way.

Even if you keep your contacts in a different contact-management application, you should be able to load them on your Kindle Fire using the techniques explained here.

Understand the Difference Between Loading and Syncing Contacts

There are two different levels of putting your contacts on your Kindle Fire:

- **Load your contacts**   In loading, you simply put your contacts on your Kindle Fire so that you can use them there. If you change your contacts on your computer or on your Kindle Fire, you'll need to find a way to propagate the changes to the other device or reconcile the differences in the contacts.
- **Sync your contacts**   In syncing, you use an application or tool that not only puts your contacts on your Kindle Fire but keeps them up to date by syncing the changes. So any changes you make to your contacts on your Kindle Fire appear on your computer as well, and vice versa.

Create vCards and Put Them on Your Kindle Fire

To load your contacts on your Kindle Fire, you create files in the vCard format. vCard is a "virtual business card" that has containers for each item of contact information—title, first name, initial, last name, job title, company, address, and so on. A vCard file can contain the information for either a single contact or for multiple contacts. When you're creating vCard files, it's usually easier to create a single vCard file that contains all the contacts you want to transfer.

After creating the vCard files, you put them on your Kindle Fire and tell the Contacts app to import them.

Create vCard Files on Your Computer

Here's how to create vCard files:

- **Windows Contacts**   Click the Start button, click your user name, and then double-click the Contacts folder to open it. Select the contacts you want to export, and then click the Export button on the toolbar (if the Export button doesn't appear, click the > > button to display it). In the Export Windows Contacts dialog box (shown here), click the vCards (Folder Of .vcf Files) item, and then click the Export button.

- **OS X Contacts or Address Book**   Open the Contacts application or the Address Book application (depending on your version of OS X). Select the contacts you want to include, and then drag them to the Desktop or a convenient Finder window. OS X creates a single vCard file with a name such as Alice Smith and 56 Others.vcf. If you want to choose the file name yourself, select the contacts, choose File | Export | vCard, and then type the name in the Save As box in the Save As dialog box before clicking the Save button.

- **Google Contacts**   In Google Contacts, select the check box for each contact you want to export, then click the More drop-down button, and click the Export item. In the Export Contacts dialog box (shown here), select the vCard Format option button in the Which Export Area, and then click the Export button. Google exports the contacts to a file named contacts.vcf in your Download folder (on Windows) or your Downloads folder (on the Mac).

Export contacts   ×

Which contacts do you want to export?
◉ Selected contacts (34)
◯ The group [My Contacts   ▾] (33)
◯ All contacts (44)

Which export format?
◉ Google CSV format (for importing into a Google account)
◯ Outlook CSV format (for importing into Outlook or another application)
◯ vCard format (for importing into Apple Address Book or another application)

[**Export**]   [Cancel]   Learn more

- **Outlook**   Click the Contacts button in the navigation bar to display your contacts. Double-click the contact to open it, and then choose File | Save As to display the Save As dialog box. In the Save As Type drop-down list, make sure the vCard Files item is selected. Choose the folder in which to save the vCard, and then click the Save button. You need to repeat this procedure for each card separately.

Put the Contacts on Your Kindle Fire

Once you've created the vCard file or files, connect your Kindle Fire to your computer and use Windows Explorer or Finder to copy the file or files to your Download folder. After doing this, tap the Disconnect button to disconnect your Kindle Fire from your computer.

 If it's easier, you can also e-mail the vCard files or files to an account on your Kindle Fire, and then save the file or files from the Email app.

After putting the file containing your contacts on your Kindle Fire, follow these steps to import the contacts:

1. Tap the Apps button to display the Apps screen.
2. Tap the Contacts icon to launch the Contacts app.

 If any of the contacts you're importing already exist in the Contacts app on your Kindle Fire, you will get duplicate contacts. The Contacts app doesn't warn you about the existing contacts or offer to merge the data—it just goes right ahead and creates new contacts.

3. Tap the Menu button to display the Menu panel.
4. Tap the Import/Export button to display the Import/Export Contacts dialog box (shown here).

Import/Export contacts

Import from internal storage

Export to internal storage

Cancel

5. Tap the Import From Internal Storage button. Your Kindle Fire searches the internal storage for vCard files and imports those it finds. The Reading vCard dialog box displays the Kindle Fire's progress, as shown here.

ⓘ Reading vCard

Reading vCard file(s)
Maria Acidda

85% 17 of 20 contacts

6. When the import finishes, the Contacts app displays the full list of contacts.

Delete a Contact

If you find you've imported a contact twice (or more times), delete the unwanted record from your Kindle Fire. Follow these steps:

1. Open each of the duplicate contacts and decide which you want to keep—usually the one with more information or newer information.
2. Open the contact you want to delete by tapping its button.
3. Tap the Menu button to display the Menu panel.

4. Tap the Delete Contact button. The Contacts app displays the Delete dialog box shown here.

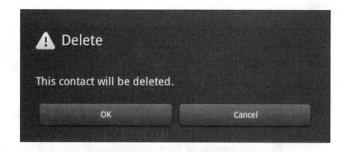

5. Tap the OK button. Contacts deletes the contact and displays the Contacts list again.

Control How Your Kindle Fire Displays Contacts

By default, the Contacts app sorts contacts by first name—Alice Zabrinski, Bob Smith, Charlene Dodds, and so on—and displays them with the first name first.

To control how the Contacts app lists the contacts, follow these steps:

1. Tap the Menu button to display the Menu panel.
2. Tap the Settings button to display the Display Options screen (shown here).

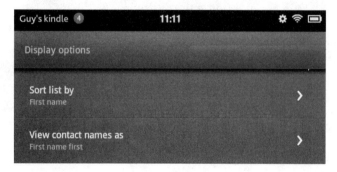

3. To change the sorting, tap the Sort List By button, and then tap the First Name option button or the Last Name option button in the Sort List By dialog box (shown here).

Sort list by

First name ●

Last name ○

Cancel

4. To change whether the first name or last name is displayed first, tap the View Contact Names As button, and then tap the First Name First option button or the Last Name First option button in the View Contact Names As dialog box (shown here).

View contact names as

First name first ●

Last name first ○

Cancel

5. Tap the Back button when you're ready to leave the Display Options screen.

Keep Your Contacts Synced Between Your Computer and Your Kindle Fire

To keep your contacts synced between your computer and your Kindle Fire, you will need to buy an application that will perform the sync for you. At this writing, these are the best options:

- **CompanionLink (Windows)** CompanionLink from CompanionLink Software, Inc. (www.companionlink.com/android/outlook/) is a program that runs on your PC and syncs with your Kindle Fire via USB. You need to install the DejaOffice app on your Kindle Fire to sync data. CompanionLink costs $49.99 to buy outright, or $14.95 for a three-month subscription.

- **Exchange By TouchDown**   Exchange By TouchDown (discussed in Project 42 for accessing your e-mail on Exchange servers) can automatically sync your contacts with your Kindle Fire.
- **SyncMate (Mac)**   SyncMate from Eltima Software (www.sync-mac.com) can sync your contacts from Contacts or Address Book and your calendar information from Calendars or iCal on your Mac with the Contacts app and the Calendar app on your Kindle Fire. SyncMate comes in a free edition that handles just the contacts and calendar, and in a pay version called SyncMate Expert ($39.99) that can also sync iTunes, iPhone, folders, and notes.

Project 45: Get Your Calendars on Your Kindle Fire

Unlike most Android devices, the Kindle Fire doesn't include a Calendar app, so if you need to keep your appointments on your Kindle Fire, you have a straightforward choice:

- **Use a browser-based calendar**   If you keep your calendars online in a service such as Google Calendars or Apple's iCloud service, you may prefer to access them through the Silk browser (or another browser). Figure 5-10 shows Silk accessing Google Calendar. Accessing your calendars through the browser is straightforward, but—obviously enough—you can use your calendars only when your Kindle Fire has a Wi-Fi connection.
- **Install a calendar app**   If you search for calendar apps in the Appstore, you'll find several free ones, such as Fliq Calendar. But if you want to sync your calendars to your Kindle Fire, you'll need to get an app that can sync the data. At this writing, this is a small field—but there may be more apps by the time you read this. If you use Google Calendar, the best app is Calengoo ($5.99), which can automatically sync your Google Calendar data to and from your Kindle Fire.
- **Exchange By TouchDown**   If your calendar is on Microsoft Exchange, the Exchange By TouchDown app can automatically sync your contacts with your Kindle Fire. See Project 42 for coverage of this app.

FIGURE 5-10 If you can rely on a
Wi-Fi connection, accessing an online
calendar through the Silk browser
is an easy way to keep up with your
appointments.

Project 46: Keep Track of Your Tasks

These days, most everyone has too much to do, and certainly too much to remember.
Your Kindle Fire can be a great way of tracking your tasks. All you have to do is install
a task-management app—and then use it.

In this project, I'll show you how to download, install, and use the task-management
app called Remember The Milk. This app is free, very easy to use, and can sync your
task list online.

If your workplace has an Exchange server and you use TouchDown For Exchange (discussed in Project 42) on your Kindle Fire, you can sync your tasks along with your e-mail.

Download, Install, and Set Up Remember The Milk

First, download and install Remember The Milk in the usual way:

1. From the Home screen, tap the Apps button to display the Apps screen.
2. Tap the Store button to display the Store screen.
3. Tap the Search box, type **remember the milk**, and then tap the Search button.
4. On the screen of search results, tap the Free button and then the Get App button.
5. When the installation finishes, tap the Open button to open Remember The Milk. You'll see the Login screen (shown here).

6. Tap the Sign Up button to display the Sign Up screen.
7. Fill out the form, and then tap the Sign Up button. You then see the Remember The Milk Home screen, which you'll meet in the next section.

Track Your Tasks with Remember The Milk

As you can see in Figure 5-11, Remember The Milk has a simple user interface featuring six main icons:

- **Today icon, Tomorrow icon, This Week icon**   Tap one of these three icons to display the tasks for today, tomorrow, or this week.
- **Lists**   Tap this icon to display the Lists screen, which you use to create and manage separate lists of tasks.
- **Tags**   Tap this icon to display the Tags screen, which you use to apply tags to tasks.
- **Locations**   Tap this icon to display the Locations screen, on which you can manage your locations.

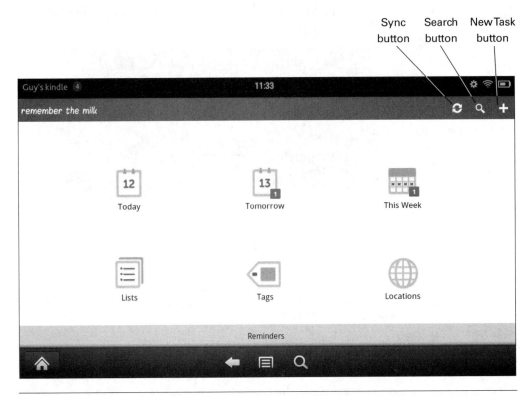

FIGURE 5-11   You can quickly access the main areas of Remember The Milk by tapping the icons on the Home screen.

To create a new task, follow these steps:

1. Tap the New Task button in the upper-right corner of the screen to display the Add Task screen (shown here).

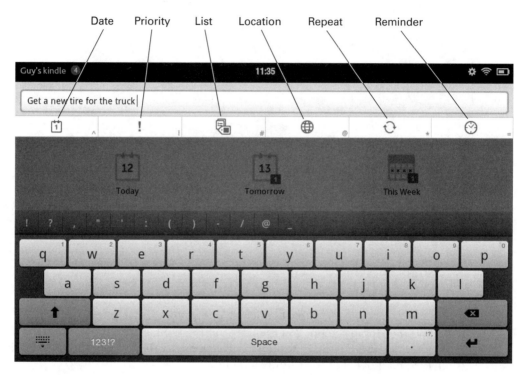

2. Type the name or description for the task in the text box.
3. Use the buttons under the text box to set the details of the task:
   - **Date**   Tap this button to display a spin wheel of days, and then tap Today, Tomorrow, or the day's name.
   - **Priority**   Tap this button to display a spin wheel showing High, Medium, and Low. Then tap the priority you want to assign.
   - **List**   Tap this button to display a spin wheel showing your task lists. Then tap the list you want to add the task to.
   - **Location**   Tap this button to display a spin wheel showing your locations. Tap the location you want to assign to the task.
   - **Repeat**   Tap this button to display a spin wheel showing repeat intervals, such as Daily, Weekly, and Biweekly. Then tap the interval you want.
   - **Reminder**   Tap this button to display a spin wheel showing reminder times, from 2 Min to 1 Hr. Tap the reminder time you want.
4. Tap the ENTER button on the keyboard to finish creating your task.

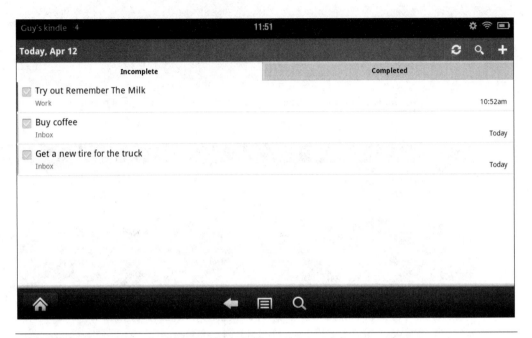

FIGURE 5-12   Tap the check box to the left of a task to mark it as complete.

To view your tasks, tap the Today button, the Tomorrow button, or the This Week button on the Home screen. You then see your list of tasks on the Incomplete tab, as shown in Figure 5-12, and can tap the check box to the left of a task to mark it as complete and move it to the Completed tab.

 To sync your tasks online, tap the Sync button on the Home screen or any of the main screens. With the basic version of Remember The Milk, you can sync only once every 24 hours. If you buy the pay version, you can sync as often as you like.

Project 47: Sync Your Notes Between Your Computer and Your Kindle

Your Kindle Fire doesn't come with a notes app, so you'll need to add one if you want to use your Kindle Fire as a notepad. The best choice is Evernote, which provides an easy way to sync your notes between your computer and your Kindle Fire—and to sync them with other devices you use, such as an iPhone or Android phone. Evernote is free software.

Even if you want to use your notes only on your Kindle Fire, not on a computer or other device, Evernote is a great app. Its automatic online sync means that your notes are safe even if you lose or destroy your Kindle Fire.

Download and Install Evernote on Your Kindle Fire

To start, download and install Evernote on your Kindle Fire. Follow these steps:

1. From the Home screen, tap the Apps button to display the Apps screen.
2. Tap the Store button to display the Store screen.
3. Tap the Search box, type **evernote**, and then tap the Search button.
4. On the screen of search results, tap the Free button and then the Get App button.
5. When the installation finishes, tap the Open button to open Evernote. You then see the opening Evernote screen (shown here). Evernote authenticates you and then displays your notes.

6. Click the Create Account button, and then follow through the process of creating an account. Or, if you already have an account, tap the Sign In button, and then type your user name (or e-mail address) and password on the Sign In screen (shown here), and then tap the Sign In button.

After you sign in, you can create notes as described in the section "Create and Sync Notes," a little later in this chapter.

Download and Install Evernote on Your Computer

If you use a computer as well as your Kindle Fire, download and install Evernote on it too.

On Windows, open your web browser and go to the Evernote website, www .evernote.com, and then click the Get Evernote, It's Free button. Click the Run button to run the installer after downloading it. After you accept the license agreement, you can either click the Install button to install Evernote with the default options or click the Advanced button to give yourself control over these options:

- Whether to install Evernote just for yourself or for all users of your computer
- Which folder to install the Evernote program files in (the default folder, Program Files\Evernote\Evernote\, is usually a sensible choice)

On Windows, Evernote runs automatically at the end of the installation.

On the Mac, click the App Store button on the Dock to launch the App Store. Type **evernote** in the Search box, and then click the Free button followed by the Install App button on the Evernote search result to download and install the application. Click the Evernote icon on the Launchpad screen or double-click the Evernote icon in the Applications folder to run Evernote.

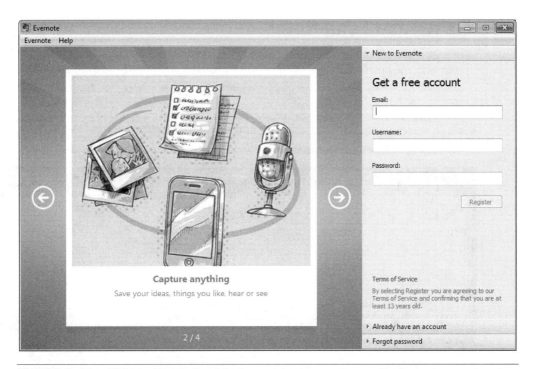

FIGURE 5-13   Click the Already Have An Account button on the opening Evernote screen, and then sign in with your user name and password.

On the opening Evernote screen (see Figure 5-13), click the Already Have An Account button and type your Evernote user name and password. On Windows, select the Stay Signed In check box if you want to stay signed in; on the Mac, select the Remember Password button if you want to store the password. Then click the Sign In button.

 If you use a smart phone (such as an Android phone or an iPhone), install Evernote on that, too, so that you're equipped to take notes and use them on whichever device you have with you.

Create and Sync Notes

After launching Evernote on your Kindle Fire, you can easily create notes. Figure 5-14 shows you the key features of the Evernote user interface, and the following list teaches you seven essential moves:

- **Create a note**   Tap the New Note button, and then work in the New Note dialog box (shown here). Type the note's title and content, and add other information as needed. When you finish, tap the Done button.

- **Delete a note**   Tap and hold the note to display the Actions dialog box, tap the Delete button, and then tap the OK button in the Delete dialog box.
- **Create a notebook**   Tap the Notebooks button on the left side of the screen, and then tap the New Notebook button in the upper-left corner (the button with a + sign on it). In the New Notebook dialog box (shown here), type the name for the new notebook, and then tap the OK button.

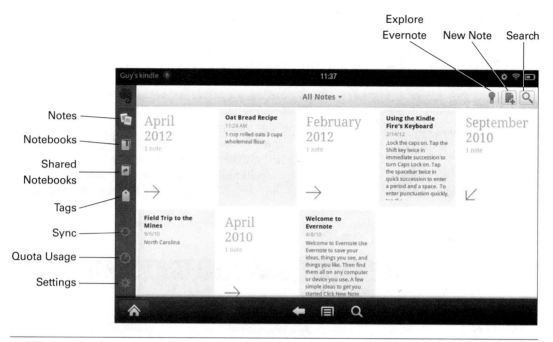

FIGURE 5-14   Evernote's user interface makes it easy to create new notes and access your existing notes.

- **Work with notebooks others are sharing with you**   Tap the Shared Notebooks button on the left side of the screen to display the Shared Notebooks pane, and then tap the notebook you want to use.
- **Change Evernote settings**   Tap the Settings button on the left side of the screen, and then work on the Evernote Settings screen that appears. When you finish choosing settings, tap the Back button.
- **Search for a note**   Tap the Search button in the upper-right corner, and then type your search term. You can search either in all your notes or tap the Advanced Search button and search only part of your data.
- **Find out more about Evernote**   Tap the Explore Evernote button to display the Explore Evernote pane, and then tap the topic you want to see.

Project 48: Send Text Messages from Your Kindle Fire

E-mail is great, but what about when you need a faster means of communication? This project shows you how to add texting to your Kindle Fire by using an app called textPlus.

Download, Install, and Set Up textPlus

The textPlus app comes in a free, ad-supported version and a pay version that costs $2.99. You'll probably want to start with the free version to see how useful you find textPlus.

Download and install textPlus from the Appstore in the usual way:

1. From the Home screen, tap the Apps button to display the Apps screen.
2. Tap the Store button to display the Store screen.
3. Tap the Search box, type **textplus**, and then tap the Search button.
4. On the screen of search results, tap the Free button and then the Get App button for the free version of textPlus, or tap the Price button and then the Buy App button for the pay version.
5. When the installation finishes, tap the Open button to open textPlus. You then see the textPlus Sign In screen (shown here).

6. Tap the Create Account button to display the Sign Up screen, and then follow through the process of creating an account.

Send and Receive Texts with textPlus

When you finish setting up your account, textPlus displays the main screen with the People tab at the front (see Figure 5-15). At the top of this tab is your free textPlus number, at which people can send you texts.

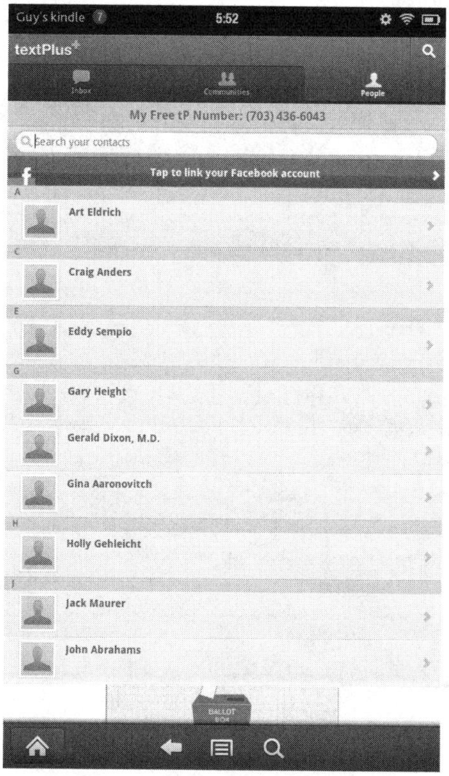

FIGURE 5-15   From the People tab in textPlus, you can quickly send texts to your contacts.

To send a text, follow these steps:

1. On the People tab, tap the recipient's name. textPlus displays the contact record.

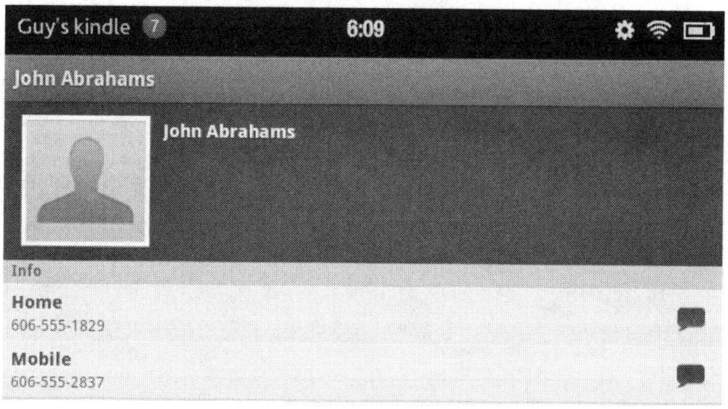

2. Tap the balloon to the right of the phone number you want to use. textPlus displays the Compose Free Message screen (shown here).

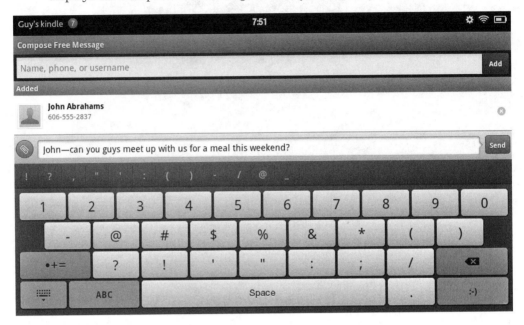

3. Type the message, and then tap the Send button.

When you receive a text message, it appears in your inbox (shown here). You can tap the message to display it on a conversation screen where you can type a reply.

Project 49: Create Office Documents on the Kindle Fire

Your Kindle Fire includes the basic version of Quickoffice, which consists of three apps—Quickword, Quicksheet, and Quickpoint—that can open Microsoft Word documents, Microsoft Excel spreadsheets, and Microsoft PowerPoint presentations, respectively. So with Quickoffice you can view Word, Excel, and PowerPoint files—but you can't edit them. Nor can you create new documents.

If you want to create new documents or edit existing ones—which, let's face it, is pretty much the point of Office apps—you need to upgrade to Quickoffice Pro, which costs $14.99 at this writing. The upgrade is easy enough—just tap the Upgrade button on the Quickoffice home screen and then follow the prompts. Figure 5-16 shows Quickpoint working on a presentation.

But before you upgrade, make sure you know your alternatives, as there are a couple of other Office suites at the same price:

- **Documents To Go** Documents To Go comes in what's called the Main App, which gives you the same level of functionality as the basic version of Quickoffice—you can open Word, Excel, and PowerPoint files, but you can't edit them or create new ones. To add the editing and creating functionality, you buy the Documents To Go Full Version Key for $14.99. Figure 5-17 shows Documents To Go with a Word document open.

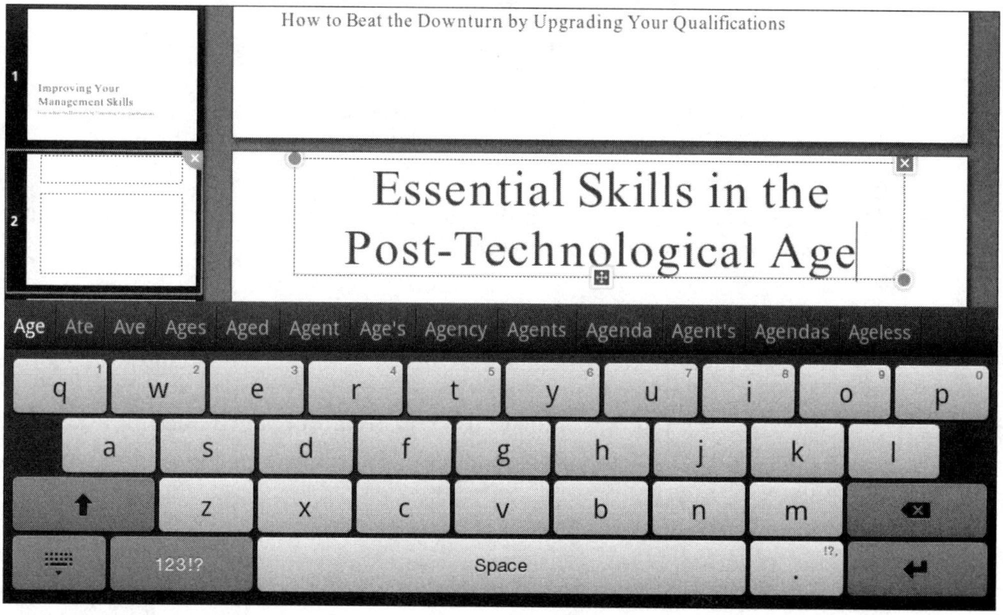

FIGURE 5-16 After you upgrade Quickoffice, you can create new Word documents with Quickword, spreadsheets with Quicksheet, and presentations with Quickpoint. You can also edit your existing Office files.

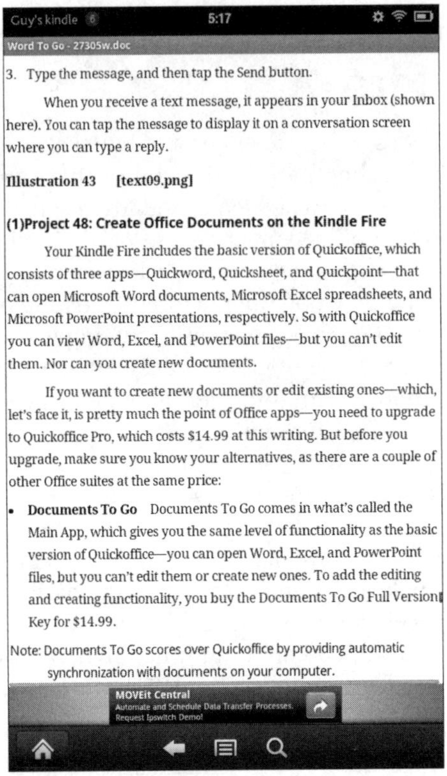

FIGURE 5-17 Documents To Go puts word processing, spreadsheets, and presentations on your Kindle Fire.

 Documents To Go scores over Quickoffice by providing automatic synchronization with documents on your computer.

- **OfficeSuite** OfficeSuite comes in two versions—the Viewer, which is free and lets you view but not edit or create Office documents, and the Pro version, which costs $14.99 and provides the editing and creating features. OfficeSuite (see Figure 5-18) makes it easy to navigate among both documents stored on your Kindle Fire and documents stored on remote accounts such as Google Docs, Dropbox, or Box.

FIGURE 5-18   OfficeSuite's interface makes it easy to navigate among your local folders and remote folders to reach the documents you need.

DOUBLE GEEKERY

Understand the Limitations of Office Suites for the Kindle Fire

Quickoffice, Documents to Go, and OfficeSuite provide an impressive amount of functionality on small devices such as your Kindle Fire or an Android phone. But these Office apps have only a small subset of the features in the full PC or Mac versions of Word, Excel, and PowerPoint.

So before you plan to start doing all your work on your Kindle Fire, take a good look at which features your chosen Office suite provides and which it doesn't—and decide whether you can do without the more powerful features.

For example, if you use Microsoft Word professionally, you almost certainly use styles to format your documents quickly and consistently. Documents To Go can apply styles, but Quickword and OfficeSuite cannot. Similarly, if you create PivotTables in Excel, or build advanced animations in PowerPoint, you will not be able to do your work on your Kindle Fire.

So assess which features you must have, and work out how well an Office suite will work for you. Generally speaking, the apps are good for jotting down ideas, creating quick spreadsheets or performing calculations by plugging figures into an existing spreadsheet, or developing the outline of a presentation. But in many cases you'll want to use a PC or Mac to perform more demanding tasks.

Project 50: Carry Your Essential Files on Your Kindle Fire

If you use your Kindle Fire extensively, you'll probably want to carry important files on it.

You can easily put the files you need on your Kindle Fire by copying them across using Windows Explorer (on Windows) or the Finder (on the Mac). But keeping more than a few files or folders up to date manually quickly becomes a chore, and it's easy to get stuck without the latest version of files.

To make sure the files on your Kindle Fire are always up to date, you can sync the files between your Kindle Fire and your computers. In this project, we'll use the free utility and service called Dropbox to sync your files.

Get and Install Dropbox, Set Up Your Account, and Log In

At this writing, Dropbox isn't available through the Amazon Appstore, so you need to side-load the app instead. You can either download the Dropbox for Android APK file from the Dropbox website or extract it from another Android device.

 See Project 34, "Side-Load Apps on Your Kindle Fire," for instructions on side-loading apps. See Project 33, "Get Apps from Other Sources than Amazon's Appstore," for instructions on extracting an APK file from an Android device. (Both projects are in Chapter 4.)

After side-loading the Dropbox APK file on your Kindle Fire, open an ES File Explorer window and tap the Dropbox APK file to launch the installation. When the installation finishes, tap the Open button to open Dropbox, and then follow through the screens that explain what Dropbox is and what it does.

When you reach the Dropbox screen (see Figure 5-19), tap the appropriate button:

- **I'm Already A Dropbox User**   Tap this button to display the login screen shown here. Type your e-mail address and password, and then tap the Log In button.

FIGURE 5-19   From this Dropbox screen, you can either log in using your existing account or start creating a new account.

- **I'm New To Dropbox**   Tap this button to display a registration screen. Type your details, tap the Register For Dropbox button, and then follow through the rest of the sign-up process. You can then log in to Dropbox.

Use Dropbox on Your Kindle Fire

Now that you've set up your Kindle Fire to use your Dropbox account, you can quickly transfer files back and forth using Dropbox. This section teaches you the moves you'll need to know.

Browse the Files and Folders You've Stored in Dropbox

To browse the files you've stored in Dropbox, tap the Dropbox tab in the upper-left corner of the screen. You see a list of the folders and files, as in Figure 5-20.

FIGURE 5-20 Tap the Dropbox tab to see a list of the folders and files you've stored in Dropbox.

From here, you can tap a folder's button to display its contents, or tap a file's button to open it in a viewer. For example, tapping the Photos folder in my Dropbox opens the Photos folder, as shown in Figure 5-21.

Tap a file's button to open it in a viewer, or tap the down-arrow button at the right end of a file's button to display a pop-up menu of actions you can take with the file, as shown here.

![Screenshot of Dropbox app showing Photos folder contents with tabs for Dropbox, Uploads, and Favorites]

FIGURE 5-21   Tap a folder's button to display its contents. You can then tap the Up To Dropbox button at the top to go back up the folder tree.

 To refresh the files in Dropbox, tap the Menu button to display the Menu panel, and then tap the Refresh button.

Save Favorite Files for Offline Access

To avoid taking up all the space on your Kindle Fire, Dropbox doesn't automatically sync all your files to the device. Instead, it makes the files available while the Kindle Fire has an Internet connection, and lets you choose which files to store on the Kindle Fire so that they're always available.

To make Dropbox store a file or folder on the Kindle Fire, you mark the file or folder as a favorite by tapping the Favorite button on the actions bar (shown in the previous illustration) or the Favorites button at the bottom of the screen when you've opened the file for viewing.

When you make a file a favorite, Dropbox downloads the file to your Kindle Fire, and then displays a blue badge containing a white star on it, as you see in the next illustration.

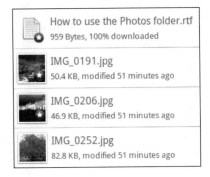

To see your full list of favorites, tap the Favorites tab at the upper-right corner of the window.

Upload a File to Dropbox

To upload a file from your Kindle Fire to Dropbox, follow these steps:

1. Tap the Uploads tab at the top of the screen to display the Uploads screen (see Figure 5-22).
2. At the bottom of the screen, tap the Other Files button to display the sdcard screen (see Figure 5-23), which shows the folders in your Kindle Fire's file system.

 On the Uploads screen, you can tap the Photos Or Videos button to display the contents of your Kindle Fire's Gallery. This is sometimes useful for sending photos, but it's really designed for Android devices that have built-in cameras. On the Kindle Fire, which has no camera, this button is that much less useful.

3. Select the check box for the file you want to upload, as shown here.

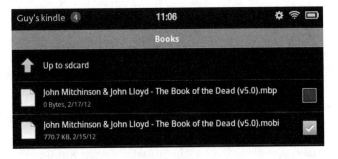

4. At the bottom of the screen, look to see which folder is selected. If it's the folder you want, leave it be and tap the Upload button. Otherwise, tap the button to display the Choose Upload Location dialog box, tap the folder, and then tap the Upload button.

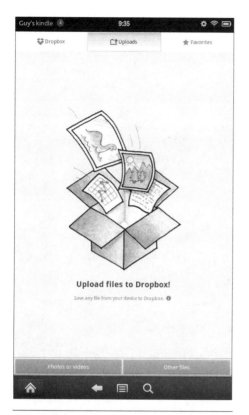

Upload files to Dropbox!

Save any file from your device to Dropbox. ❶

| Photos or videos | Other files |

FIGURE 5-22   On the Uploads screen, tap the Other Files button to upload any other kind of file.

sdcard

Android

Audible

Books

Documents

download

Evernote

kindleupdates

LOST.DIR

Music

Pictures

pulse

TunnyBrowser

Video

Select files to upload

Dropbox

| Cancel | Upload |

FIGURE 5-23   On the sdcard screen, navigate to the folder that contains the file you want to upload, and then select its check box. Select the check box for each file you want to upload, as shown here.

5. Dropbox then uploads the file, showing a readout of its progress, as you see here.

Guy's kindle   11:08

⚙ 📶 🔋

📦 Dropbox   📤 Uploads   ⭐ Favorites

📄 John Mitchinson & John Lloyd - The Book of the Dead (v5.0).mobi

6. When you finish uploading files, tap the Dropbox tab to return to the list of files in your Dropbox.

Index

Numbers